UNDER THE MICROSCOPE

UNDER THE

JEREMY BURGESS

MICHAEL MARTEN

ROSEMARY TAYLOR

MICROSCOPE

A HIDDEN WORLD REVEALED

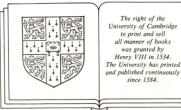

*The right of the
University of Cambridge
to print and sell
all manner of books
was granted by
Henry VIII in 1534.
The University has printed
and published continuously
since 1584.*

CAMBRIDGE UNIVERSITY PRESS

CAMBRIDGE

NEW YORK   PORT CHESTER

MELBOURNE   SYDNEY

Published by the Press Syndicate of the University of Cambridge
The Pitt Building, Trumpington Street, Cambridge CB2 1RP
40 West 20th Street, New York, NY 10011, USA
10 Stamford Road, Oakleigh, Melbourne 3166, Australia

First published under the title *Microcosmos* 1987
First paperback edition as *Under the Microscope* 1990
Printed in the United States of America

British Library cataloguing in publication data

Burgess, Jeremy
Under the Microscope.
I. Microscopy
I. Title.   II Marten, Michael, 1947–.
III. Taylor, Rosemary, 1952–.   IV. Microcosmos.
502.82

Library of Congress cataloguing in publication data applied for

ISBN 0 521 39940 8

Text: Jeremy Burgess, Michael Marten, Rosemary Taylor,
        Mike McNamee, Rob Stepney
Editing: Michael Marten, Rosemary Taylor
Design: Richard Adams/AdCo Associates
Diagrams: Neil Hyslop

Special thanks to: Donald Claugher, Gordon Leedale, Lou Macchi,
Barry Richards, Cath Wadforth, Derek Wight

CONTENTS

1. MICROCOSMOS Jeremy Burgess & Michael Marten   6

2. HUMAN BODY Rob Stepney   12
   Reproduction   14
   Vision   18
   Hearing   20
   Taste   22
   Nervous system   24
   Muscle   28
   Bone   30
   Respiration   32
   Digestion   34
   Blood   36
   Immune System   39
   Skin & Hair   40

3. ANIMALS Rosemary Taylor & Michael Marten   42
   Protozoa   44
   Parasitic Worms   48
   Rotifers   52
   Insects   54
   Compound Eyes   60
   Caterpillar Hatchery   62
   Mites   64

4. SEED PLANTS Jeremy Burgess   66
   Roots   68
   Stem & Wood   70
   Leaves   72
   Attack & Defence   76
   Flowers   78
   Pollination   82
   Embryo   84
   Seeds   86

5. MICRO-ORGANISMS Jeremy Burgess   88
   Viruses   90
   Bacteriophages   92
   Bacteria   94
   E. Coli   96
   Rhizobium   98
   Algae   100
   Diatoms   102
   Fungi   104

6. THE CELL Jeremy Burgess   108
   Nucleus   111
   Endomembrane System   112
   Mitochondria   114
   Chloroplasts   116
   Cytoskeleton   118
   Specialised Cells   119
   Mitosis   122

7. INORGANIC WORLD Mike McNamee   124
   Atoms   126
   Dislocations & Grain Structures   128
   Heat-treatment Structures   130
   Dendritic Structures   131
   Crystal Structures   134
   Petrology   138
   Diagenesis   140
   Ferrous Metals   142
   Non-ferrous Metals   144

8. INDUSTRIAL WORLD Mike McNamee   148
   Material Joining   150
   Solid Phase Joining   152
   Extreme Duty Materials   154
   Ceramics   156
   Bioengineering   158
   Electronics   160
   Corrosion   166
   Coating   167
   Failure Analysis   168
   Quantitative Microscopy   172

9. EVERYDAY WORLD Jeremy Burgess   174
   Fabrics   176
   Velcro   178
   Paper   180
   Watch   180
   Records & Discs   182
   Food   184

TECHNICAL APPENDIX Jeremy Burgess   186
   History of Microscopy   186
   Light Microscopy   190
   Electron Microscopy   196
   Other Types of Microscopy   201

CREDITS   206

INDEX   208

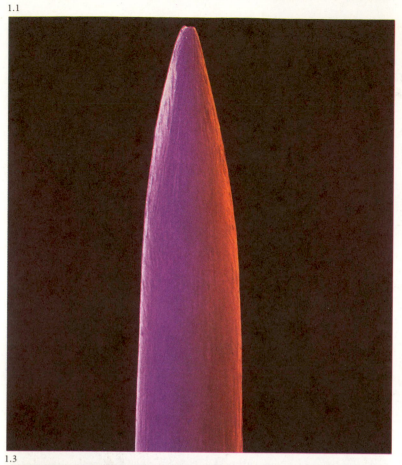

1.1

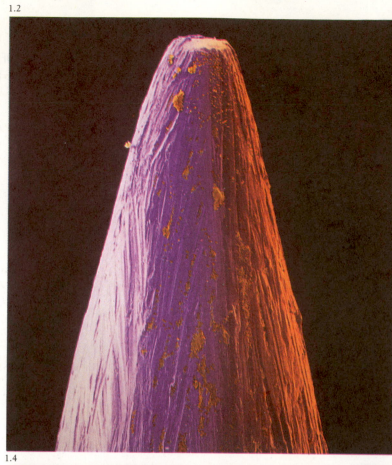

1.2

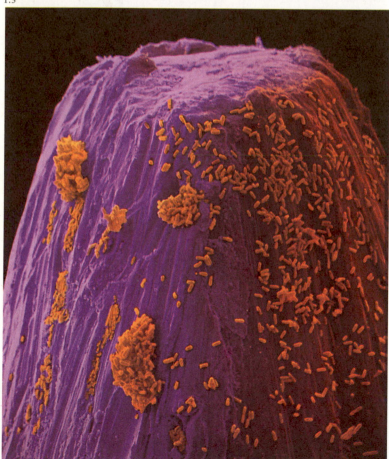

1.3

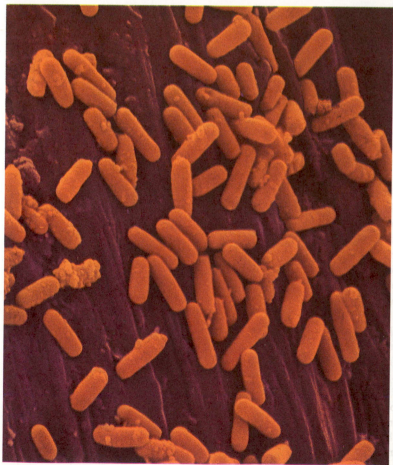

1.4

6 MICROCOSMOS

Microscopes do not just magnify; because of their resolving power, they also reveal fine detail. As magnification is increased, successively finer detail is revealed until the microscope's resolution limit is reached and nothing new can be distinguished.

Many micrographs are easy to understand. This is the case with the 'realistic' pictures produced by the SEM and many of the pictures taken through light microscopes. Other micrographs are more difficult to interpret and their full appreciation requires some understanding of how the specimen has been prepared and how the magnified image of the specimen is produced.

In a light microscope, magnification is the result of light travelling from the specimen through two glass lenses, the objective lens and the eyepiece. The light may reach the objective after passing through the specimen (transmitted light), or after being reflected from its surface (incident light). Transmitted light can only be used if the specimen is naturally translucent or if it consists of a very thin section which allows light to pass through it. Most of the light micrographs in this book, particularly the biological ones, were made using transmitted light. Incident light is used when the specimen is a whole object, or naturally opaque – a microchip, for instance, or a piece of metal.

In a TEM, magnification is the result of a beam of electrons travelling down through the centre of a series of circular electromagnets, called 'electron lenses'. In one sense, the TEM resembles the light microscope – the electron beam is transmitted through the specimen. But the electrons must travel in a vacuum, and the method by which the image is formed is not analogous to that which occurs in a light microscope. Electrons are not very penetrating particles, and TEM specimens must therefore be very thin indeed to enable the electron beam to pass through them. The typical TEM specimen is about 1/1000th of the thickness of a single page of this book. This requirement determines the main characteristic of transmission electron micrographs: they represent only a tiny section of a whole object.

The SEM can be likened to an optical microscope used in incident light mode. A very fine beam of electrons, again travelling in a vacuum, is focused by electron lenses onto the specimen. As the beam strikes it, other electrons are emitted from the specimen's surface and radiate outwards. These 'secondary electrons' are collected by a detector and used to produce a point of light on a television screen. The image is built up by scanning the electron beam over the specimen in a series of lines and frames, called a raster scan. As the electron beam moves over the specimen, it produces a series of points of light on the television screen, so that a complete image is built up. The magnification is a direct result of the ratio of the specimen area scanned to the area of the television screen. If the electron beam is set to scan the whole of a microchip, the magnification might be of the order of ×50; if it scans just a few of the microchip's multitude of circuits, the magnification might be ×10 000.

Because the secondary electrons are emitted from its surface layers, the SEM specimen does not need to be a thin section. It can be a whole object, up to a size determined by the microscope's specimen chamber. Specialist SEMs have been built for the examination of objects as large as a whole knee joint and even a gun barrel. Secondary electrons are not the only particles produced when the primary beam strikes the specimen surface. X-ray photons are also emitted, and so occasionally are photons of visible light. And some of the primary electrons may be 'back-scattered', or reflected, off the specimen. These radiations can also be used to form an image. X-ray imaging is particularly useful, because the X-rays are emitted with energies characteristic of the elements that make up the specimen's surface. It is therefore possible, for example, to map the positions of the different elements in an alloy or composite material (see Figure 8.40).

Before it becomes a specimen for microscopy, an object must usually be prepared in some way. For the SEM this preparation may amount to no more than coating the object with an extremely thin layer of gold to improve its surface conductivity and emission of secondary electrons. Biological SEM specimens may also need a simple freezing step to stabilise them against the vacuum inside the instrument. Many light microscope and TEM specimens, on the other hand, are thin sections and require elaborate and skilled preparation.

Sections make it possible to view opaque objects and examine their internal structure. And in the case of both the light microscope and the TEM, the thinner the section, the higher the potential resolution of the image. Methods have therefore been devised for slicing or grinding objects of all kinds until they are wafer-thin. In the case of light microscope specimens, such sections are usually about 5 micrometres (1/200th of a millimetre) thick. TEM specimens need to be about 200 times thinner at 100 nanometres (1/10 000th of a millimetre) or less.

1.7 In order to produce a clear image, specimens for the SEM are usually coated with a metal such as gold to increase their ability to reflect electrons and their electrical conductivity. The photograph shows a microchip being covered with a gold layer only 5–10 nanometres thick by a process called sputter-coating. The specimen is in a vacuum chamber containing argon gas at low pressure. A flat plate of pure gold is positioned a few centimetres above the microchip, and a voltage of about 1000 volts is applied between it and the gold plate. As a result, the gas ionises, and argon ions strike the gold plate so forcefully that gold atoms are ionised and dislodged from it. These gold ions then stream downwards and form a uniform layer on the surface of the specimen. The ion plasma is seen in the picture as the bright glow; its shape is controlled by a circular magnet around the gold plate at the top of the picture.

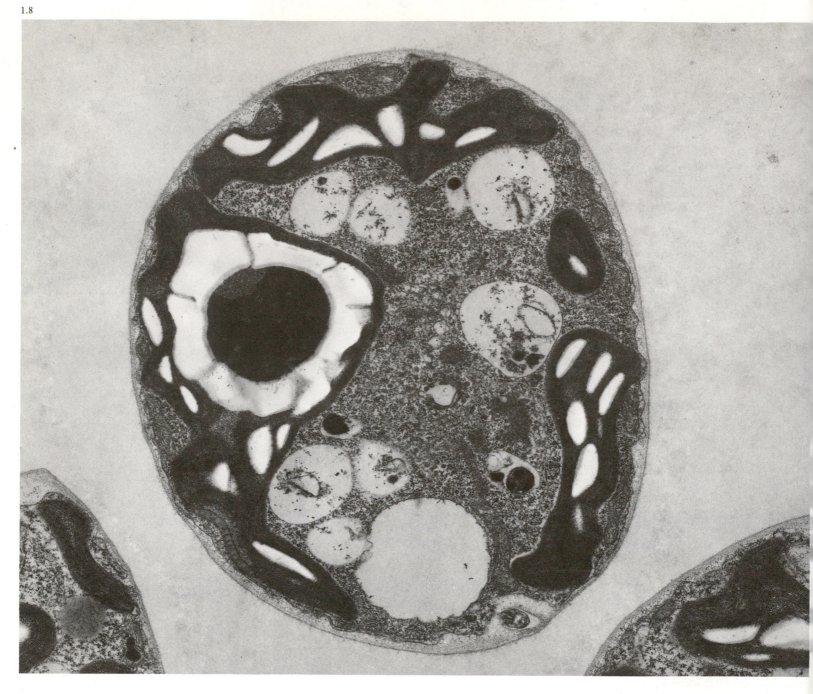

1.8 The problems of image interpretation are at their most acute in the case of thin sections photographed using a TEM. This picture typifies such micrographs. It is a section of the unicellular green alga, *Chlamydomonas asymmetrica*, at ×20 000 magnification. In looking at this image, as with all sections, it is important to remember that what appears in the picture is a two-dimensional representation of a tiny part of a three-dimensional object. The cell is about 10 micrometres in diameter; the section represented by this picture is only 70 nanometres thick, or rather less than 1 per cent of the thickness of the whole cell.

Because the section is two-dimensional, the cell appears round although it is spherical, and the pure white areas, which are sections of ellipsoidal starch grains, appear elliptical. Furthermore, not all the components of the cell are visible in this tiny sampling of it; there is no nucleus in the picture, for example, although the cell would certainly contain one. The dark areas around the periphery of the cell are sections through chloroplasts. There are five distinct dark areas visible, although two nearly touch. It is possible that all these areas are in fact part of a single chloroplast which is sufficiently

undulating to disappear from view in such a thin section. Likewise, the circular grey regions in the cytoplasm, the vacuoles, may indeed be discrete spherical spaces; on the other hand, some of them at least may coalesce in a plane which is not included in this section. Microscopists habitually examine large numbers of sections to establish three-dimensional relationships when this is important. Equally, they habitually select sections for photography which are as 'perfect' as possible, and fairly describe the 'typical' structure of their object. This picture is free from technical defects such as staining dirt, scratch lines from

a faulty microtome knife, and folds. It is a good 'typical' picture of *C. asymmetrica* insofar as it clearly shows the pyrenoid in the chloroplast (the black area within the white ring). This was why it was taken. It is a poor picture of an algal cell, because it does not show the nucleus; however, it is highly probable that the dozen or so sections which would give a view of the nucleus would in fact not contain the pyrenoid in this particular cell. In this sense, micrographs are created by their photographers; they do not happen on their own.

TEM, stained section, ×20 000

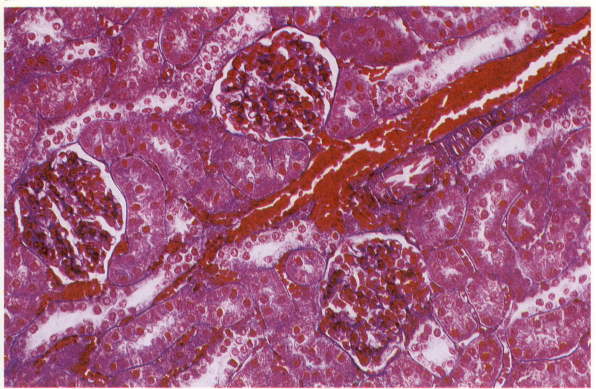

1.9 The thin section of human kidney tissue in this light micrograph contains a blood vessel, which runs diagonally from top right towards bottom left, and three glomeruli. Our kidneys have thousands of glomeruli, each of which is a tiny bundle of tightly packed capillaries which filters the blood of excess water and poisonous substances. The bright colouring of the tissue results from the stain used during the preparation of the specimen.
LM, trichrome stain, ×940

Biological thin sections are usually transparent and so low in contrast that they need to be stained in order to produce a satisfactory image. Different stains can emphasise particular features in a specimen. For example, staining a section of a cell with uranium acetate, a heavy metal salt, makes the genetic material – the DNA and RNA – appear black in a transmission electron micrograph. In light microscopy of human and animal tissue, a dye called eosin is widely used and colours the cytoplasm of cells various shades of pink.

Since colour is a property of light, the image produced by a light microscope is seen in colour, although it may be photographed in black-and-white. The colours themselves may be due to staining, or to the form of illumination chosen by the microscopist, or they may be the natural colours of the specimen. Natural coloration is common in pictures of small living organisms, but even here the operator can decide whether the creature is seen against a black background (dark field illumination) or a light one (bright field illumination). And in the former case, filters can be used to tint the background any chosen colour (Rheinberg illumination).

Other illumination techniques change the colour of a specimen completely. The use of polarised light, for example, produces colours by the process of birefringence. The technique can produce very colourful images of crystalline materials, including common chemicals like vitamin C or aspirin, and this has made it popular with many amateur microscopists. It is also used quantitatively. The colours displayed by a section of rock of known thickness, for example, result from fundamental properties of the minerals from which the rock is composed. A geologist can therefore use the technique as a precise tool for identifying rock and mineral types.

Electron micrographs are always originally monochromatic. Unlike photons of visible light, electrons do not 'carry' colour and therefore can tell us nothing about a specimen's coloration. But electron micrographs can be artificially coloured by computer, photographic, or hand-tinting techniques. Such 'false colour' can make it easier to distinguish particular structures in a specimen, but it is usually added for aesthetic effect and it need bear little or no relation to the natural colour of the object or organism portrayed.

The 300 or so pictures in this book represent only a tiny sample of modern microscopy. We have not attempted to include examples of every technique or every specialist microscope; nor have we tried to cover every application of microscopy. The aim has been to produce a book of some of the best and most informative micrographs, and through them to introduce the microcosmos.

(The figure captions in this book all end with a 'technical line' in small type. This gives information about the type of microscope used to produce the image, and the magnification. Also included where relevant and known are details about the illumination technique, the type of section and the stain. All scanning electron micrographs in the book have been produced by secondary electron imaging except where otherwise stated. Abbreviations used in the technical lines are: LM – light microscope; SEM – scanning electron microscope; TEM – transmission electron microscope; STEM – scanning transmission electron microscope; HREM – high-resolution electron microscope; DIC – differential interference contrast; H & E – haematoxylin and eosin.)

CHAPTER 2
THE BODY

OVER the past three centuries, microscopy and medicine have advanced hand in hand. They continue to do so. Because structure and function are intimately linked, progress in our understanding of body processes in health and disease has to a considerable extent depended on our ability to see increasingly fine structural detail in our organs and tissues. And it is advances in our understanding rather than new equipment or drugs that underlie the most important developments in medicine.

The 16th century brought the first systematic dissection of animal corpses, and growing interest in the *post mortem* examination of the human body. From what was readily visible with the unaided eye, physicians began to relate disease to the gross appearance of the internal organs. In the 17th century microscopists started to probe the hidden structure of biological materials. When Leeuwenhoek died in 1723, he bequeathed 26 microscopes. With each microscope came an example of the specimens for which it had been built: among them were blood, sperm, hair and muscle.

With an increasing use of microscopes in the late 18th century came an interest in texture, and the realisation that organs are composed of various forms of 'tissue' – a word that first came into use at this stage. The basic material of which we are composed was thought to be fibre, at times dense, and at others loosely woven.

In the mid-19th century, the greater resolving power of microscopes revealed the cell as the basic building block of tissue and the site of the disordered processes underlying disease. Progress was also dependent on advances in the preparation of specimens, particularly in the cutting of thin sections, and in the increasing use of dyes to reveal structure by staining. In the 1870s and 1880s, almost every coloured substance known was tried as a stain in microscopy; and the advent of synthetic dyes greatly added to the limited range of plant extracts available.

By the end of the 19th century microscopists had realised that a dark-staining nucleus was found in almost all cells, and that it was perpetuated when cells divided. They had seen ribbon-like chromosomes within it, and by 1900 had directly observed the process of fertilisation. Only with these observations of the fundamental features of cells did it become clear how male and female make equal contributions to their offspring; and how the characteristics of one generation are transmitted in an orderly way to the next.

More recently, the advent of electron microscopy has revolutionised our concept of the cell, revealing complicated structures within it that account for many of its specialised functions. It was not until transmission electron microscopy of muscle cells in the 1950s, for example, that we learned how muscle contraction works.

But medical microscopy is above all a working, practical technology. With most cancers, for example, symptoms and gross appearance are only an uncertain guide. Examination by microscope is still the way malignancies are confirmed. And surgeons often hang fire in the operating theatre until the potentially abnormal specimens they have taken can be examined. This is the province of histology – the study of normal tissue – and of histopathology – the study of its diseased counterpart.

The process of preparing a histological specimen for light microscopy begins by taking tissue representative of the feature under study, and fixing it (often in formalin) to prevent putrefaction and degeneration. The sample is then placed in increasing concentrations of alcohol to remove all water. Once dehydration is complete, the alcohol is removed by immersion in an organic solvent such as xylene. The next stage is to embed the tissue in a medium that provides strength and support. In light microscopy, it is usual to impregnate the specimen with paraffin wax. The solidified block of tissue is then cut by a microtome. This machine shaves off a series of sections, each around 4–5 micrometres thick, forming a ribbon of samples. The ribbon is floated in water, and sections picked up on a glass microscope slide, ready to be dried and stained.

The use of stains is crucial in histology: the history of the science is one of improved staining techniques as much as of improved microscopes. Haematoxylin and eosin are stains in routine use. All specimens are first treated using these two dyes. Their effect is to stain the nuclei of cells blue and the cytoplasm shades of pink. A battery of dozens of stains can then be brought to bear to pick out individual structures.

Given the immense diversity of tissues within the body, and the range of microscopical techniques available to study them, the images in this chapter cannot provide a comprehensive coverage. And although microscopy is an essential tool in pathology, the body as it appears in these pages is predominantly a healthy one. The pictures are also mostly of human specimens. However, where structure and function are essentially the same, images of tissue from other animals – usually other mammals – are included.

2.1 This light micrograph shows a wafer-thin section through one of the folds of the wall of the jejunum – the second part of the small intestine. The area covered is about half a millimetre across. The intestine at this point is deeply folded and lined with many thousands of finger-like projections, or villi, which protrude into the hollow tube along which food passes. Around 20 villi can be seen in this picture. Villi are all of much the same length, but the unusual angle of this section makes them appear of differing size. Each villus is composed of a mucous membrane, here stained pink, and a central core of blood vessels and small lymphatic channels. The mucous membrane is made up largely of columnar cells, called enterocytes. It is just possible to distinguish the individual columnar cells aligned in a single row to form the surface of each villus. The stain used in the preparation of the specimen has dyed the nucleus at the base of each cell a dark red. Enterocytes are involved in the breakdown of food, but their major role is in the absorption of its constituent nutrients. Scattered among the enterocytes are goblet cells, which have stained pale blue. Goblet cells secrete the mucus lining that protects the intestine from self-digestion. Nutrients absorbed by the enterocytes pass into the rich network of blood capillaries and lymphatic vessels at the centre of each villus. Extending along the base of the villi is the *muscularis mucosae*, containing blue-stained smooth muscle cells.
LM, trichrome stain, ×430

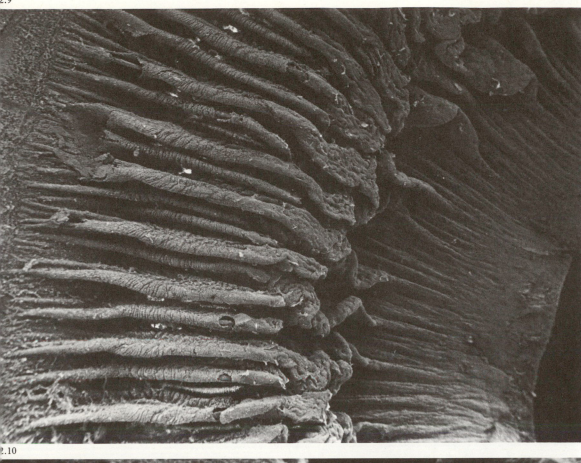

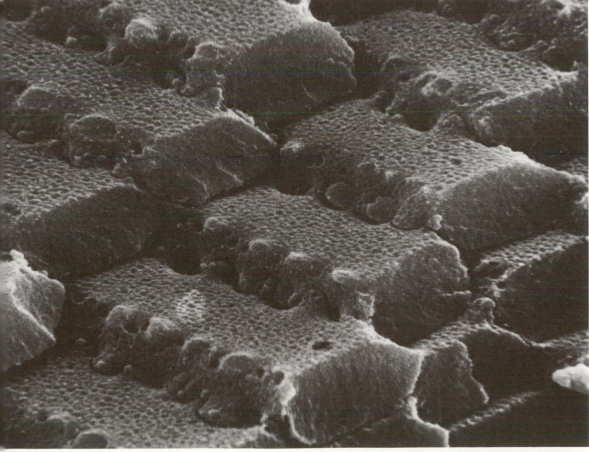

2.8 This micrograph and accompanying diagram show a cross-section through the complex sandwich of cells that makes up the mammalian retina. The front of the retina is at the bottom of the picture, so light would enter from below and strike the photosensitive rod receptors (the neat row of cells at the top of the image) only after it has first passed through several layers of cell bodies. The bleaching of light-sensitive pigment contained in the outer segments of the rods generates patterns of electrical activity. These signals are transmitted via branching nerve endings to bipolar nerve cells, which pass the information on through junctions with ganglion cells that eventually form the fibres of the optic nerve. Large, irregularly-shaped Müller cells extend through the depth of the retina, providing support and nutrition.
SEM, ×4350

2.9 The deeply-folded tissue that occupies most of this low-magnification image is called the ciliary body. It normally encircles and controls the lens of the eye, flattening it to focus distant objects onto the retina and making it more spherical to accommodate those that are closer. In this view, seen from the back of the eye, the lens has been removed. As a result, we can see through to the less-folded muscles of the iris, which control the aperture of the pupil (the black area at bottom right) and so the amount of light that reaches the retina.
SEM, ×15

2.10 The remarkable transparency of the 4-millimetre thick lens of the eye is due to the absence of nuclei in its cells, and to the crystalline precision with which they are arranged. The zipper-like rows of ball-and-socket joints that lock long lines of cells together may also play a part. In cross-section, lens cells are flattened hexagons, arranged in regular stacks. The length of lens cells, 10 millimetres, is about 2000 times their thickness, 5 *micro*metres. So it is not surprising that they are usually thought of as fibres.
SEM, ×6240

HEARING

Our organ of hearing – the cochlea
– is structured like a snail's shell.
It is a tapering, fluid-filled spiral
cavity 3.5 centimetres long.
Membranes running the length of
the spiral divide it into three
parallel ducts, the upper and lower
ducts being connected at the apex.

2.11 In this light micrograph, the coils
of the cochlea are seen in section.
The surrounding bone appears blue
and the fluid-filled chambers of the
spiral are white. The two membranes
are clearly seen. The upper membrane
appears tissue-thin; the lower – the
basilar membrane – is more
substantial. The three ducts they form
are also easily distinguished.

Positioned in the middle duct – on
the basilar membrane that forms its
floor – is the organ of Corti. Here, in
cross-section, it appears as a small
wedge shape in each of the cochlea's
coils. Rather like a spiral piano
keyboard, the organ of Corti runs the
full length of the middle duct. It is this
part of the cochlea that actually
responds to sound. At its top is a shelf-
like projection – the tectorial
membrane – and below it rows of
sensory cells rooted in the basilar
membrane. These 'hair cells' – seen in
more detail in the scanning electron
micrographs on the opposite page – are
ultimately responsible for our
perception of pitch and intensity.

Sound waves striking the ear drum
are transmitted via three small bones
(the ossicles) to an opening, called the
oval window, at the start of the upper
duct of the cochlea. There the
vibration is translated into movement
in the cochlear fluid. This movement
causes the basilar and tectorial
membranes to rise and fall, producing
shearing forces that stimulate electrical
activity in the hair cells lying between
them. Fibres from the hair cells
bundle together to form the auditory
nerve, which leads to the brain.

The loudness of a perceived sound
seems to depend on the *magnitude* of
vibration in the basilar membrane.
Intense noises produce a large
displacement, and generate a fast rate
of impulses from the hair cells.

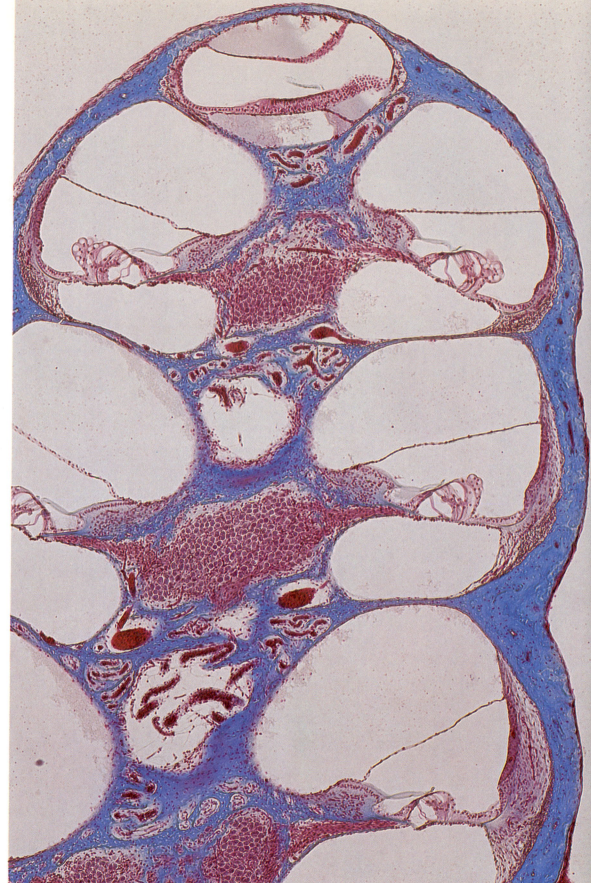

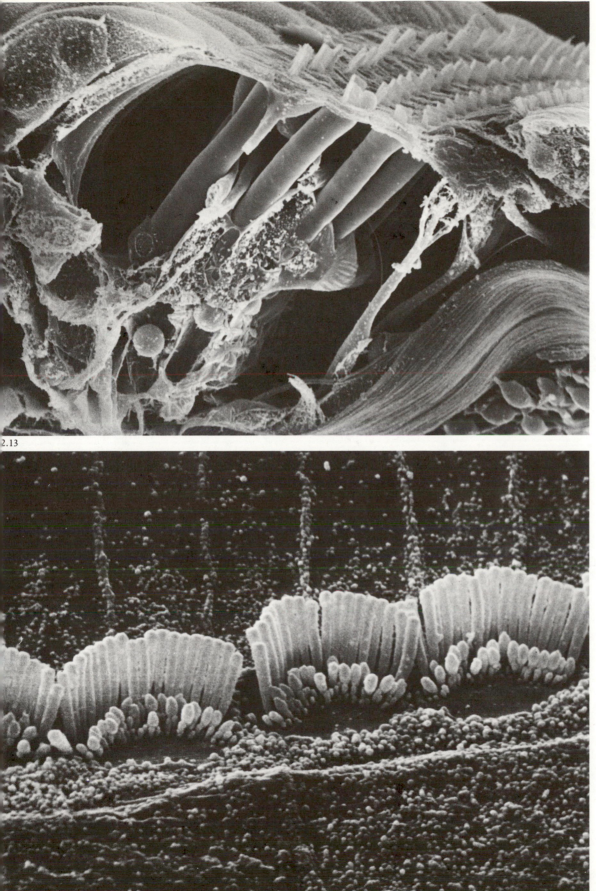

Perception of pitch is more complicated, but seems to involve the *position* along the basilar membrane at which cell firing is greatest. We know that hair cells in different regions of the cochlea are sensitive to different frequencies. High-frequency sounds have a short wavelength and generate a movement in the cochlear fluid which peaks near the broad origin of the spiral tube, exciting maximally the hair cells at that point. With lower tones, the wave amplitude takes longer to build up, reaching its peak near the tapering end of the tube, and producing the greatest stimulation among a different population of cells. Apparently, the organ of Corti's greatest sensitivity corresponds to the frequency of the human scream.
LM, trichrome stain, ×234

2.12 More than 20 000 hair cells, each having as many as 100 separate hairs, make up the organ of Corti. Their task is to translate mechanical movements caused by their displacement into electrical impulses. Towards the top of this micrograph are four rows of hairs, three of them stoutly supported by the pillar-like Deiter cells below. The basilar membrane has become buckled during preparation of the specimen and appears as à wavy sheet at bottom right. Normally, the tectorial membrane would lie above the hairs, making contact with the tallest of them.
SEM, ×1970.

2.13 At greater magnification, the hair cells are seen to produce several neatly ordered rows of hairs, which are also called stereocilia. The tall hairs are clearly arranged according to height. The smooth area just in front of the smaller stereocilia is also part of the hair cell surface.
SEM, ×7585

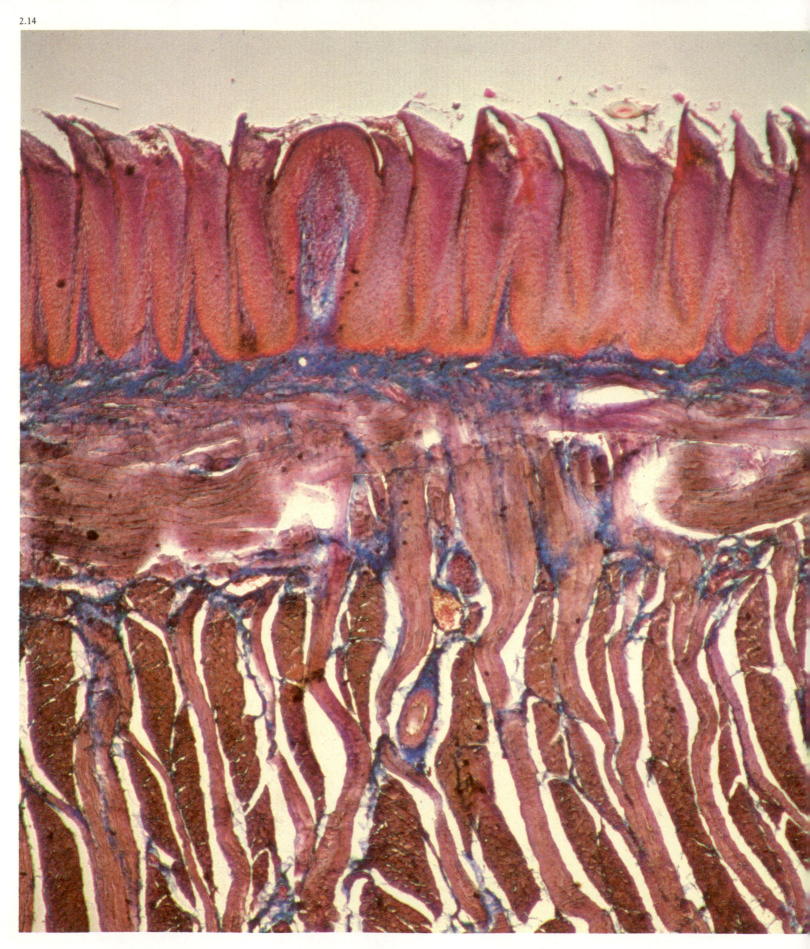

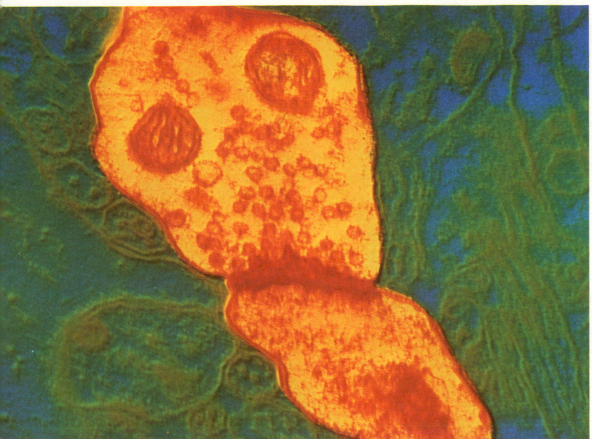

2.25

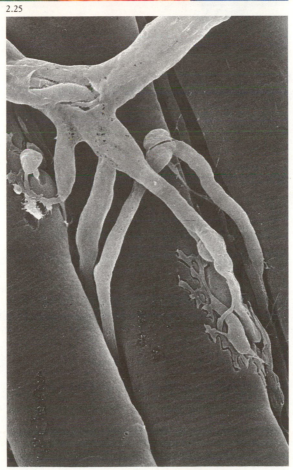

2.22 The outer zone of the cerebellum forms the upper two-thirds of this micrograph. Below it are four spider-like Purkinje cells, which are among the largest neurones in the body. The cerebellum receives input from areas of the brain responsible for initiating movement, and from the body's sense receptors, but we do not know precisely how its cells integrate motor and sensory information and coordinate fine movement.
LM, silver stain ×980

2.23 This electron micrograph shows a junction, or synapse, between two neurones of the human cerebral cortex. Although the knob-like ending of the upper nerve seems to touch the post-synaptic membrane of the one below, there is a minute gap between them, which is coloured red. Transmission of information along nerves is by electrical impulse. But its passage from one nerve to another involves neurotransmitter chemicals. The small red and orange specks in the upper nerve ending are sac-like vesicles containing neurotransmitter. When activated, the biochemicals are released to diffuse across the synaptic cleft. They then interact with receptors on the next neurone, triggering or inhibiting a new electrical impulse. The two larger circles towards the top of the image are mitochondria – organelles that produce energy for the cell.
TEM, false colour, ×78 300

2.24–2.25 A motor neurone activates muscle cells through end-plates – small swellings that form the terminals of the axon. These neuromuscular junctions are portrayed in a light micrograph (Figure 2.24) and by scanning electron microscope (Figure 2.25). Where fine motor movement is required, one neurone may control one muscle fibre. But for cruder control, as here, a single axon can innervate several muscle fibres. The passage of instructions from nerve to muscle relies on the same chemical processes as synaptic nerve–nerve transmission.
2.24 LM, gold chloride impregnation stain, ×170
2.25 SEM, magnification unknown

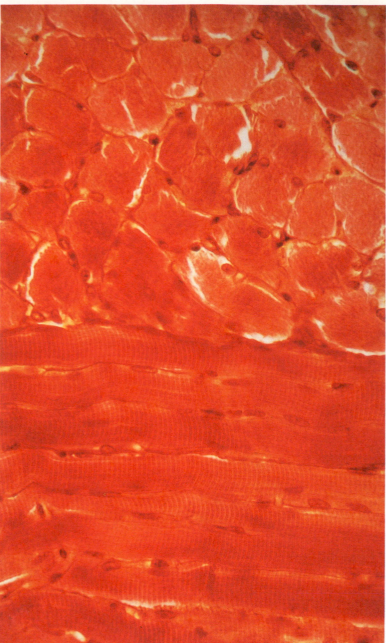

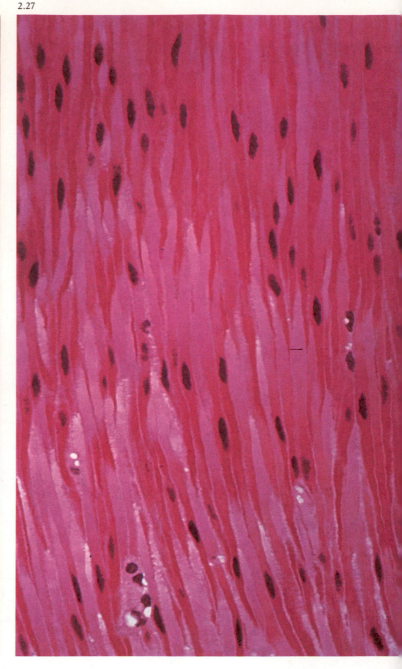

MUSCLE

Muscle cells form three distinct kinds of tissue. The most familiar, because it moves our limbs, is skeletal muscle. It is also called striated muscle, because of its prominent banding. This banding is absent in so-called smooth muscle, which performs largely automatic processes such as the contraction of blood vessels and of the gut. Similar to skeletal muscle in appearance, but unique in its capabilities, is cardiac muscle, which is capable of setting up its own rhythm of contraction, independent of nervous system control. It is unique too in its tirelessness – contracting every second for a lifetime.

2.26 Human striated muscle is seen here both in cross-section (top) and longitudinal section (bottom). Skeletal muscle cells are long – hence their description as muscle *fibres*. Each fibre has several nuclei spaced at intervals along it. Both parts of this micrograph show that these dark-staining nuclei are positioned at the edge of the cell. Running the length of each fibre, barely visible in the longitudinal part of the section, are contractile proteins arranged in thin threads called myofibrils. More prominent is the banding *across* the fibres, caused by myofibrils being composed of alternating light and dark sections. LM, ×620

2.27 The spindle-shaped cells that form smooth muscle are grouped in irregular bundles. They do not generally have to contract with the same force as skeletal muscle, nor with the same speed or precision. There is only one nucleus per cell. LM, H & E stain, ×445

2.28 The small black dots in this cardiac muscle cell are granules of animal starch, or glycogen. Glycogen is chemically broken down to glucose, and glucose in turn yields energy to power muscle contraction. Most of this chemistry takes place in mitochondria – the elliptical structures at left and bottom right. The picture also shows clearly the long threads of the myofibrils and their alternating light and dark sections.
TEM, stained section, magnification unknown

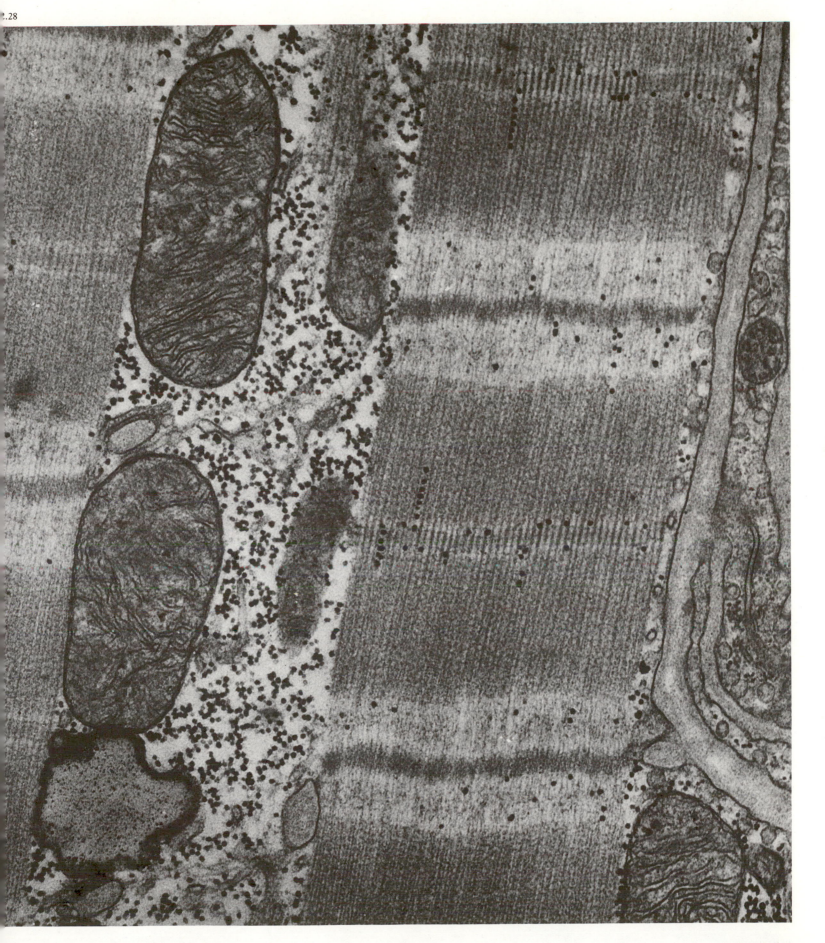

BONE

The human skeleton gives the body shape and support. It is made up of 206 rigid bones, together with the cartilage (gristle) that lines the joints and forms the cushion-like discs separating the vertebrae. The whole structure is supremely well engineered. In typical long bones, such as the femur, compact bone forms the solid outer wall of the shaft, while the interior is composed of more spongy material. This gives a high resistance to mechanical stress for a low overall weight. Bones also anticipated efficient architectural structures such as the dome (the skull), column (the femur, humerus) and arch (the foot).

As well as providing form, the skeleton plays a crucial role in maintaining the body's internal biochemical environment – acting as a store of minerals for the body to call on in times of need. In their hollow centres, bones also house marrow, which acts as the production line for both white and red blood cells.

The hardness of bone derives from its high content of calcium salts. But bone is far from inert, for all its mineral strength. Depending on the demands of the body, it is constantly growing or being resorbed. The calcium is laid down in a matrix called osteoid, which is secreted by 'bone-building' cells, or osteoblasts. These cells encircle canals that carry blood vessels.

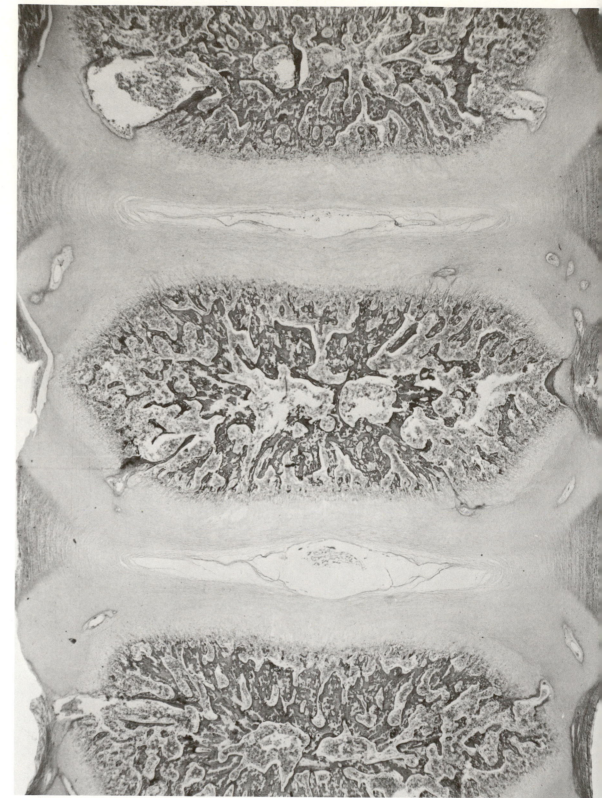

2.29 In this low-magnification light micrograph of the human backbone, the tissue in the centre of each vertebra appears sponge-like in section. Between the three areas of bone are two intervertebral discs – in effect hydraulic shock absorbers that protect the spine and brain from unnecessary jolting. The discs are made mostly of cartilage strengthened with collagen fibres, which are flexible but extremely strong and do not stretch. These fibrocartilage discs enable the vertebrae to be connected to each other in a way that allows some degree of movement, without any sacrifice in strength. The fine concentric layers of cartilage, which appear light grey, surround a white space – the *nucleus pulposus* – filled with a thick fluid. It is this unusual form of liquid connective tissue that acts as the hydraulic element in the shock absorber. Discs do not actually slip out from between the bones of the spinal column; but with advancing age there is a tendency for the ring of cartilage to weaken, allowing the nucleus pulposus to be extruded. This produces the painful condition known as a slipped disc. LM, ×18

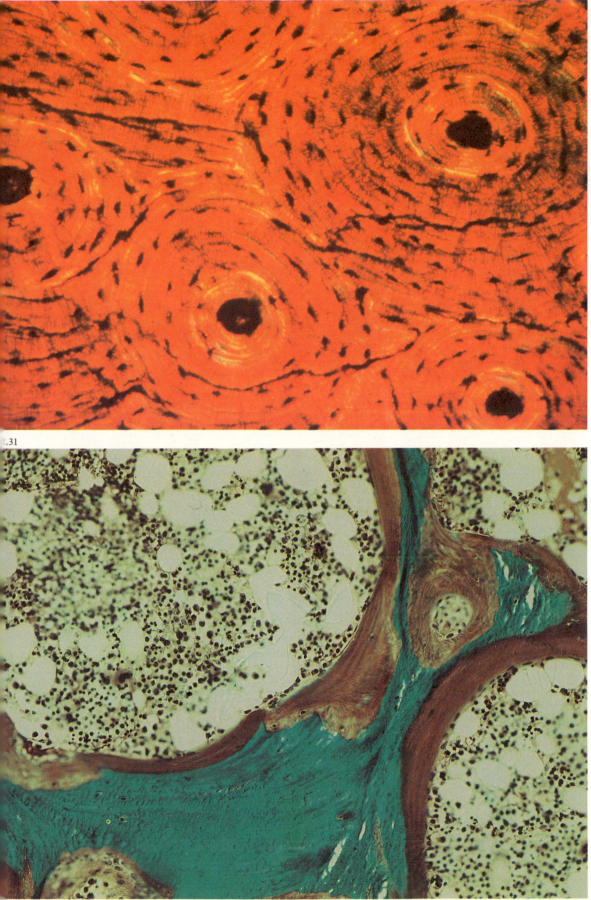

2.30 The shaft of long bones such as the femur is composed of compact bone, seen in cross-section in this micrograph. Concentric circles of bony material surround large channels (black) that contain blood vessels, lymph vessels and nerves. These channels, called Haversian canals, run the length of our bones. Looking like lines of stitching around each canal are lacunae – literally holes in the bone. Each one is occupied by an osteocyte (an osteoblast cell that has become embedded within the bone matrix). Each osteocyte sends out fine projections to make contact with its neighbours, giving the cells a spider-like appearance. Haversian canals start out wide, and gradually become filled in as layers of bone are formed. Each canal, together with its surrounding bony plates, forms a Haversian system. Haversian systems follow the lines of stress within a bone, acting like scaffolding poles. Living bone is constantly being remodelled. Large osteo*clast* cells break it down, and its components are removed. As this happens, old Haversian systems disappear and new ones come into being.
LM, Schmorl's picro thionin stain, ×210

2.31 The appropriate use of stains can reveal osteomalacia, or 'softening of the bones', a condition known as rickets when it occurs in children. If uncorrected, it leads to abnormal bending of the bones and deformity in adult life. Osteomalacia is caused by a deficit of vitamin D, usually because of poor diet and inadequate exposure of the skin to sunlight. It also occurs in clinical conditions such as chronic kidney failure. The problem is that osteoblast cells continue to secrete their organic matrix, osteoid, but deposition of calcium crystals is delayed. There is therefore accumulation of unmineralised osteoid tissue. In this micrograph, active osteoblasts appear as the tiny dark granules, and fully mineralised bone is green. Unmineralised osteoid appears as the brown layer in between.
LM, undecalcified resin section, Goldner's trichrome stain, magnification unknown

RESPIRATION

The respiratory system exists to enable oxygen from the atmosphere to be absorbed by the blood, and to allow carbon dioxide to be excreted in return. Once absorbed, oxygen is conveyed to the tissues, where it is needed so that carbohydrate fuel can be burned to provide energy. The respiratory tract is essentially a system of branching tubes, made largely of cartilage, which convey air from the mouth and nose to the sites in the lungs where gas exchange takes place. Although

our lungs can contain 5 litres of air when fully expanded, each breath typically holds 0.5 litres.

The respiratory tract proper begins with a single tube, the trachea, which then forks to form the right and left main bronchi. Each bronchus in turn divides into two or three smaller tubes, and then into innumerable finer airways called bronchioles. Because exchange of gases requires a large surface area, bronchioles end in clusters of tiny sacs, or alveoli. The presence of so many cavities gives a healthy lung the appearance of an air-filled sponge.

The body's tube-like organs – such as the gut and the respiratory tract – are lined with tissue rather like the skin that covers our exterior. This lining, called

epithelium, consists of layers of closely-packed cells with little cementing substance in between. Beneath the epithelium is a supporting layer of connective tissue, usually containing glands. The epithelium and this layer form the mucosa.

2.32 Clumps of fine hairs, called cilia, protrude from the tops of specialised epithelial cells lining the bronchial passages. Along with the cilia, this scanning electron micrograph shows numerous 'goblet cells'. Their function is to release mucus onto the surface of the epithelium. Rhythmic movement of the cilia serves to move bacteria and other particles away from the gas-exchanging parts of the lung and towards the throat where they can be swallowed or coughed up.
SEM, ×950

2.33 In contrast to the healthy state in the previous image, this micrograph shows a cancer of the bronchus – the most common form of lung cancer. Such cancers are caused by smoking and are almost always fatal. They frequently occur near the point where the trachea forks to form the two bronchi, since this area suffers from heavy deposition of the carcinogenic tars contained in tobacco smoke. The disorganised region of malignant cells at bottom right is invading the normal ciliated epithelium at left and top. Cancers consist of primitive cells which have not developed any specialised function. In this instance, for example, no cilia are present. The prime characteristic of cancer cells is their uncontrolled growth.
SEM, ×950

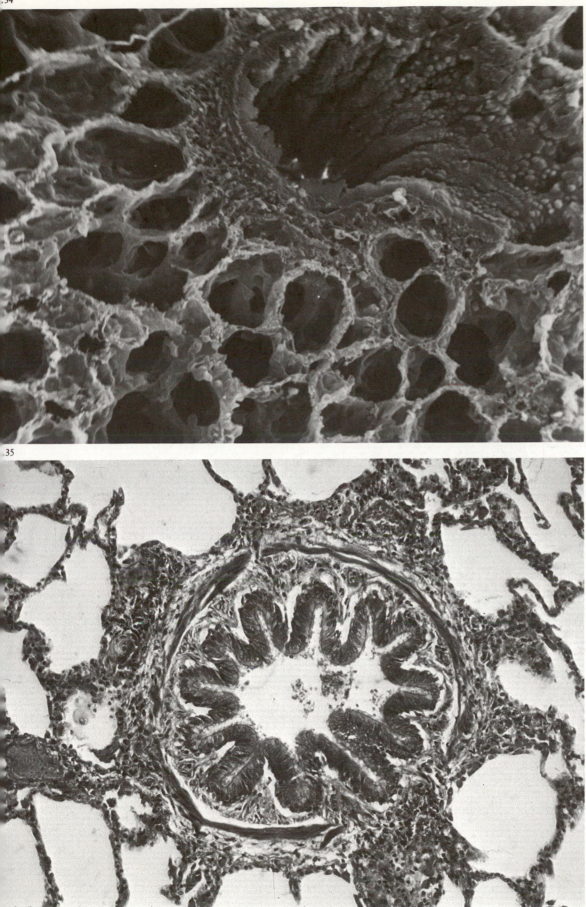

2.34 The fine lacework walls of the alveoli, and a single bronchiole (top), are seen in this scanning electron micrograph. The human lungs contain around 700 million alveoli, giving a total surface area the size of a tennis court. The walls of the alveoli are extremely thin, usually being composed of just a single layer of cells. Immediately next to them (although not visible in this specimen) are fine blood capillaries whose walls consist of a single layer of cells. This means that the barrier between air on the one side and blood on the other is only 0.3 micrometres thick; oxygen and carbon dioxide readily diffuse across. As the conducting airways reach into the depths of the lungs, they become increasingly narrow; and their walls are composed more of smooth muscle and less of cartilage. The inner surface of the bronchiole shown here appears corrugated. The structure of its highly folded epithelial lining is seen more clearly in the light micrograph below. SEM, ×280

2.35 This micrograph shows almost the same view as the previous scanning electron micrograph, but here the bronchiole is seen in cross-section, with the white areas of the alveoli around it. At 0.5–1.0 millimetre in diameter, bronchioles are the smallest of the lungs' airways. The deeply folded surface of the bronchiolar epithelium is clear. It is made up of simple rows of column-like cells, surrounded by a ring of smooth muscle cells which controls the bronchiole's diameter. Abnormal contraction of bronchiolar smooth muscle, as in asthma, severely restricts the flow of air and gives rise to symptoms of breathlessness and wheezing. Beyond the encircling smooth muscle are the alveoli. Their thin walls are composed of flattened epithelial cells and more rounded cells which are thought to secrete surfactant, a detergent-like material that lowers surface tension and improves the elasticity of the lungs. LM, H & E stain, ×180

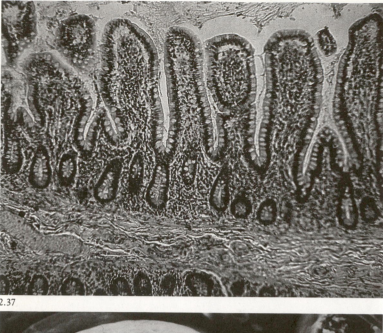

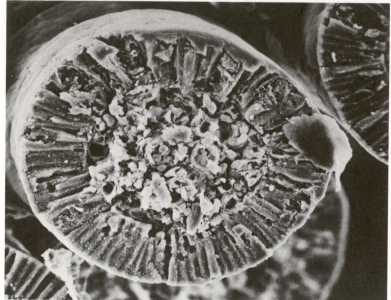

2.37

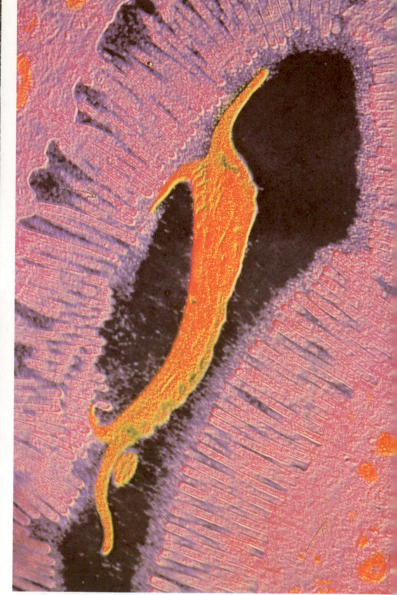

DIGESTION

Before food can be used by an organism, it must first be broken down. The gastrointestinal tract, or gut, is the site of both the digestion and absorption of food. Digestion is achieved primarily by means of enzymes, proteins which catalyse specific chemical reactions. It occurs mainly in the stomach and the upper part of the small intestine, the duodenum. Enzymes are secreted into the intestinal contents by glands in the gut wall, and by separate organs, such as the pancreas. Since this part of the gut is geared to the chemical breakdown of tissue, care must be taken that it does not digest itself. Protection is normally achieved by the secretion of mucus onto the intestinal surface. Mucus also eases the passage of food. In addition to chemically digesting material, the gut is designed to mix, pound and move it along.

Digestion breaks food down into molecules small enough to be absorbed through the gut wall. This uptake of nutrients occurs mostly in the jejunum (the second part of the small intestine), where the surface area available for absorption is enormously increased by the presence of countless villi – promontories of tissue, like loading wharves, that protrude from the gut wall.

2.36 This section through the human intestinal wall shows it lined by finger-like villi. The 'skin' of the 'fingers' is formed mostly by columnar cells, called enterocytes, with a dark nucleus at their base. These cells have a fast turnover. Formed by rapid cell division in 'crypts' at the bottom of the pits between the villi, enterocytes progress up their sides, only to be shed 4 days later. Tissue full of blood vessels forms the core of the villi, and runs along their base. The products of digestion are absorbed into this network of vessels, and subsequently transported throughout the body. LM, haematoxylin and van Gieson stain, ×10

2.37 This electron micrograph provides a view complementary to that of Figure 2.36, showing how a single intestinal villus is constructed. The column-like cells forming its outer surface are interspersed with mucus-secreting goblet cells which, in this specimen, are concentrated in the top half of the villus and are distinguished by the granular secretory droplets they contain. Large blood vessels and smaller capillaries make up the core. SEM, ×615

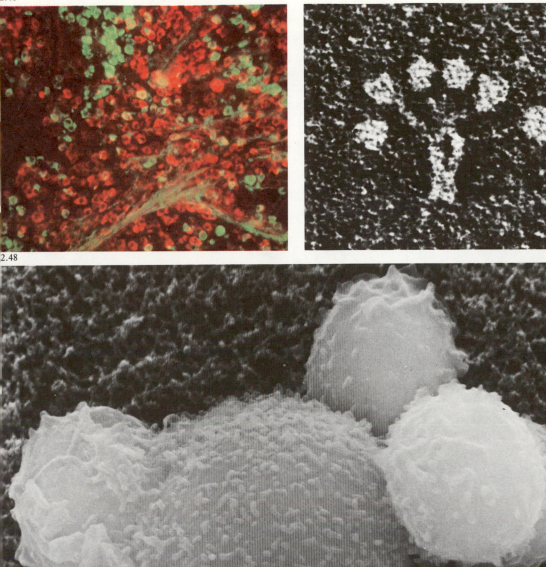

2.48

plasma cells in a mouse spleen are producing two classes of antibody protein: those stained red are secreting immunoglobulin G (IgG), and those stained green immunoglobulin M (IgM). The fact that no cell is stained both red and green shows the specificity of the immune response. Cells specialise in the production of antibodies of one particular sort. The image was obtained by staining tissue from the mouse spleen with two fluorescent dyes. Antibodies to IgM were labelled with the green dye fluorescein, and antibodies to IgG with the red dye rhodamine. The labelled antibodies then bound selectively to cells carrying the corresponding immunoglobulin on their surface.
LM, fluorescence, ×950

2.47 When an antibody combines with its antigen, it can trigger the destruction of the cell to which the antigen belongs in a number of ways. One is by activation of 'complement' – a system of nine proteins, designated C1–C9, which circulate in blood plasma. The tree-shaped protein molecule in this extremely high-magnification transmission electron micrograph is termed C1q. An antibody attached to a target cell will bind to one of C1q's six 'heads'. This triggers a cascade of the other complement proteins, from C1 to C9, which causes the punching of holes in the membrane of the target cell, leading to its destruction.
TEM, negative stain, ×1 400 000

2.48 As well as being activated by invading organisms, the immune system responds to any of the body's own cells that are abnormal, such as tumour cells. This scanning electron micrograph shows four small T-lymphocytes (so called because they mature in the thymus gland) attacking a large cancer cell. Some T-cells produce substances that attract patrolling macrophages and stimulate their phagocytic activity. Others – like these – attack the target cell directly and are known as killer T-lymphocytes. Unfortunately, as we know from the prevalence of cancers, their efforts are not always successful.
SEM, ×1125

IMMUNE SYSTEM

Bacteria and other organisms that penetrate our outer defences contain large proteins (antigens) which the body recognises as foreign. One way we respond is by producing antibodies – specific proteins which circulate in the blood until they can combine with their target antigen. Manufacturing antibodies is the job of B-lymphocytes. Lymphocytes are white cells which mature in tissues of the lymphatic system such as the thymus and spleen. Other kinds of lymphocyte – the T-cells – act more directly to destroy invading organisms.

2.46 B-lymphocytes that encounter antigens divide to form antibody-secreting 'plasma' cells. Such cells are capable of secreting 2000 identical antibody molecules per second for the few days of their mature life. In this immunofluorescence micrograph,

SKIN & HAIR

Although only a few millimetres thick, our 25 square metres of skin form the body's largest organ, and an important interface with the world. Because it is only sparsely covered with hair, human skin is vulnerable. But the 'horny', cornified, outer layer – the epidermis – is some protection; and when intact the skin is an effective barrier against viruses and bacteria. Below the epidermis is a dense layer of tissue – the dermis – rich with blood vessels and nerve endings. The skin is a versatile sensory organ, responsive to touch, pressure, pain, heat and cold; and vital in temperature regulation.

2.49 Pronounced valleys and ridges are found in skin from the palm of the hand (shown here), the soles of the feet and the fingertips. The precise pattern is unique to each of us. Sweat pores appear as small craters along the ridges. Cells nearest the skin surface – the cornified layer – have been flattened and hardened by deposition within them of the protein keratin. This tough, dead barrier of cell remnants is continuously being shed, and its flakiness is evident in the micrograph. Dead cells are constantly replaced by newer, maturing cells, which take a month to migrate from the base of the epidermis.
SEM, ×30

2.50 Like the opening to a deep cave, a sweat pore spirals down through the outer layer of the skin. It ends in a coiled sweat gland in the dermis or subcutaneous layers. This pore is on the palm of the hand – one of the areas of the body richest in sweat glands. One quarter of the body's heat loss is achieved through sweating.
SEM, ×560

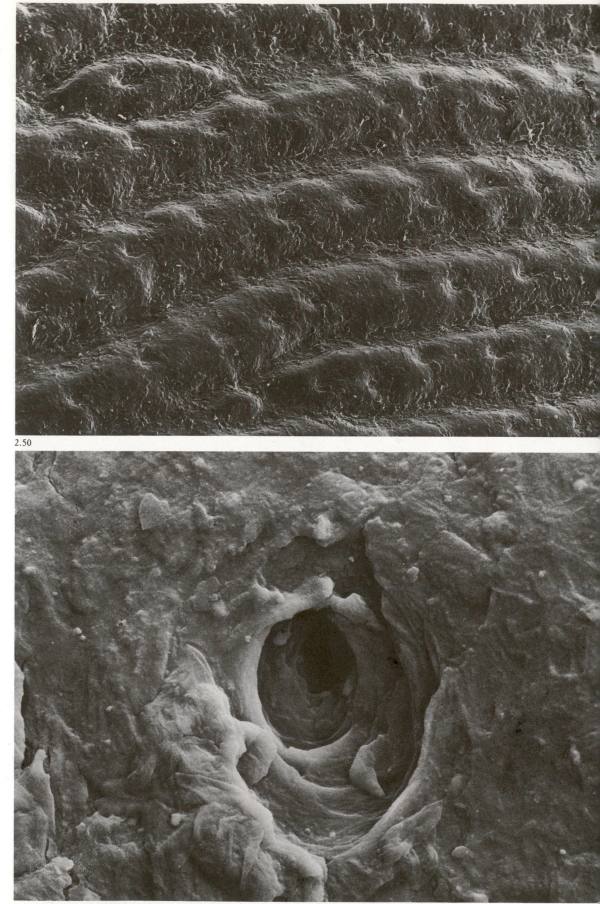

2.49

2.50

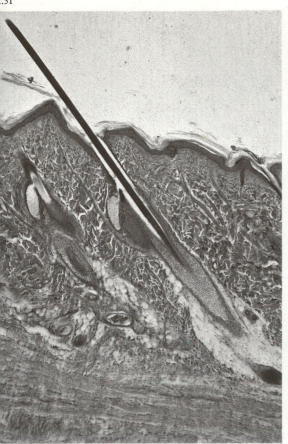

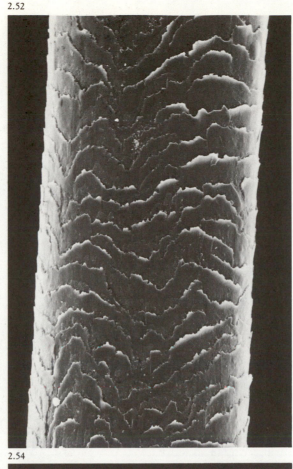

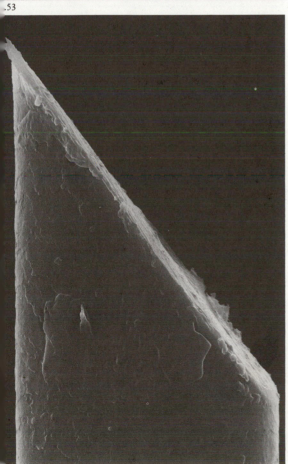

2.51 The different layers of the skin, and a single hair, are shown in section in this light micrograph. The horny outer layer of the epidermis is seen as the thin, darker band, supported by the much deeper but less dense dermis, which merges into the subcutaneous layer at the bottom of the picture. Note how the epidermis grows down to form the lining of the hair shaft, or follicle, below the skin surface. The hair is rooted in an enlarged portion – the hair bulb, a region of actively dividing cells from which the hair grows. Hair may curl when the follicles are curled, or when the hair bulb lies at an angle to the shaft. Associated with each follicle is a bundle of smooth muscle, responsible for pulling the hair erect in conditions of cold or fear, and one or more sebaceous glands that secrete the oily, waterproofing agent sebum onto the hair and the skin surface. A sebaceous gland can be seen half way down the left side of the hair follicle.
LM, magnification unknown

2.52 Hair consists largely of keratin. This scanning electron micrograph of the surface of a normal human hair clearly shows the overlapping keratin plates, or scales, which are thought to reduce hair matting.
SEM, ×480

2.53–2.54 This is the kind of comparison that infuriates the manufacturers of electric razors. On the left is a man's beard hair, cleanly sliced by the blade of a 'wet' razor. On the right is a beard hair from the same man, torn and mangled by the action of an electric razor. Since each hair is approximately one fifth of a millimetre in diameter, the difference visible to the naked eye is little or none, and the difference felt by someone kissing is likely to be more imagined than real. Nonetheless, one can see why the manufacturers of electric razors would have preferred it if the scanning electron microscope had never been invented.
SEMs, both ×275

CHAPTER 3
ANIMAL LIFE

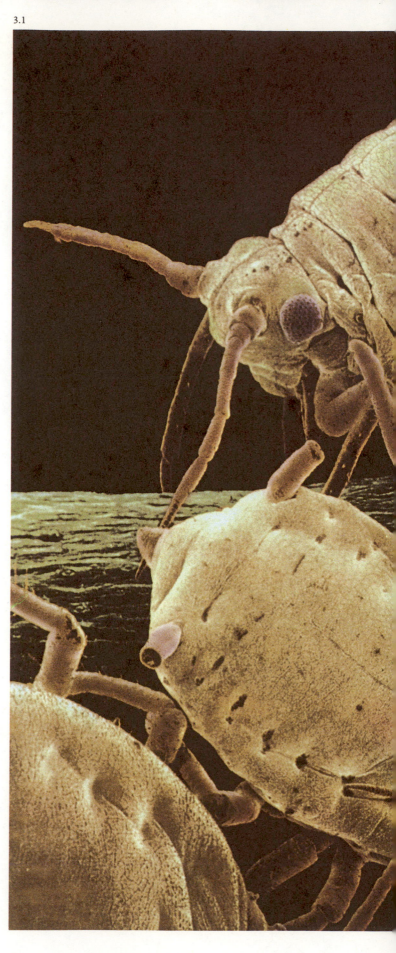

MICROSCOPES serve to counteract our biased view of the animal kingdom. The mammals, birds, reptiles, amphibia and fish which loom so large in our consciousness disappear altogether. They are replaced by the protozoa, worms and arthropods covered in this chapter. In terms of both the number of species and the number of individuals, it is these creatures which truly dominate the animal life of Earth.

The most important characteristic defining a creature as an animal is the way in which it obtains energy for growth. Most plants and many micro-organisms need only simple minerals, water, and usually air and light. Animals require preformed food from an organic source. This requirement dictates many aspects of their lifestyle, such as the development of sensory systems and locomotion to enable them to catch prey and avoid becoming prey themselves. Feeding and digestive mechanisms are adapted to the food source. Insects have a great variety of chewing, cutting, or piercing mouthparts for eating wood, leaves, or blood. Tapeworms have no mouth and no digestive system; they inhabit the intestines of other animals, where they absorb the already digested food of their hosts. The degree to which the body of an animal has specialised in directions like these provides a basis for classification.

The most 'primitive' animals have relatively poor senses and a simple body structure consisting of just one cell, or a colony of single cells joined together. Protozoa nonetheless perform all the functions of an animal within their single cell, which can be extremely complex. 'Higher' animals have a multicellular body with a nervous system and specialised organs and tissues; the cells which comprise them usually perform one or a handful of specific tasks, and may be comparatively simple.

Important steps in the evolution of 'higher' animals were the development of a special internal body cavity or 'coelom' (in the roundworms); the development of separate body segments (in the annelid worms); and the acquisition of a protective cuticle or exoskeleton (in the arthropods). Taken together, they made possible the great range of structures and lifestyles of the arthropods in general and the insects – the most successful of all animals – in particular.

Amateur and professional light microscopists have long delighted in observing small aquatic animals, from protozoa to the larvae of insects and crustacea. Many move with great speed and elegance, seeking out food, escaping danger, or reproducing as one watches. The scanning electron microscope cannot show us a living world, but its detailed images of small animals reveal how complex they can be.

3.1 The aphids (Aphidoidea) are a large, successful group of insects. Commonly known as plant lice, they are specialised plant feeders that exploit two different hosts during a season. This group of unidentified wingless aphids is feeding on a plant stem. The proboscis of the aphid at far right has penetrated the stem and is drinking the sugary juices within. Large populations of aphids develop rapidly and plants are quickly depleted of food, causing them to wilt and die. At the posterior end of each aphid are two short tubes, called cornicles, which secrete pheromones and wax. At the front end are the segmented antennae, positioned in front of the lateral compound eyes.
SEM, false colour, ×175

PROTOZOA

The protozoa, or 'first animals', are by definition located at the start of the animal kingdom. They are its simplest representatives, and the vast majority consist of a single cell with all the equipment necessary for an animal existence. A few species classed as protozoa are exceptional, in that they obtain energy from sunlight in the same manner as plants. Some 80 000 species have been described and these are divided into four classes: the flagellates, amoebas, ciliates and sporozoa.

Protozoa have not only the basic cellular organelles such as a nucleus and mitochondria, but also an array of additional structures that enable them to feed, excrete waste, protect themselves and move around the environment. Some species form an external protective layer called a pellicle, and some a complex shell or test.

Most protozoa feed like animals by breaking down organic compounds and absorbing their constituents. Many have a 'cell mouth', which among the ciliates is an elaborate, fixed structure, while in some amoebas it is temporary and disappears once feeding ceases. As the food reaches the animal's cytoplasm, it is enclosed in a food vacuole, where it is digested by the action of enzymes.

Several mechanisms are used by protozoa for moving about their environment. The flagellates use flagellae – whip-like filaments that cause an undulating, sometimes gyratory, motion through the water. Ciliates rely on cilia, or fine hairs, which beat against the water, creating a current. Amoebas advance using pseudopodia, or 'false feet'. These are extensions of their cytoplasm, which they channel in one or

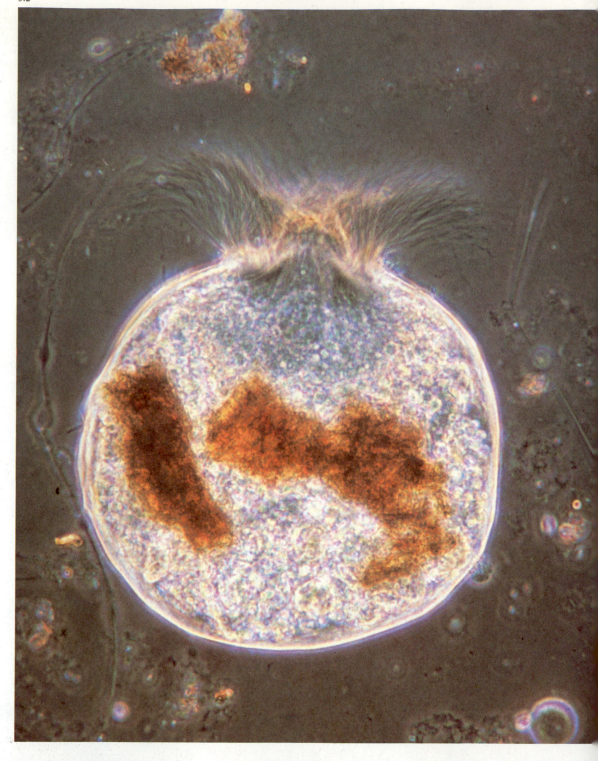

several directions simultaneously.

Protozoa are found in freshwater, seawater and in abundance in the soil. The majority form a valuble link at the base of the food chain. Some species, however, are parasitic and a few of these cause serious disease in humans. The malarial parasite *Plasmodium* is a protozoon, and so is *Entamoeba hystolytica*, the cause of amoebic dysentery.

3.2 The flagellate *Barbulanympha ufalula* forms a symbiotic relationship with a rare wood-eating cockroach, *Crytocercus puntulatum*. It lives in the hindgut of the cockroach, where it provides the essential digestive enzymes lacking in its host. The cockroach feeds exclusively on wood, but is unable to convert cellulose into carbohydrate. In return for doing this *B. ufalula* shares in the digested products. The cytoplasm contains two brown fragments of half-digested food above which is the multiflagellate crown, concealing the mouth area. LM, ×520

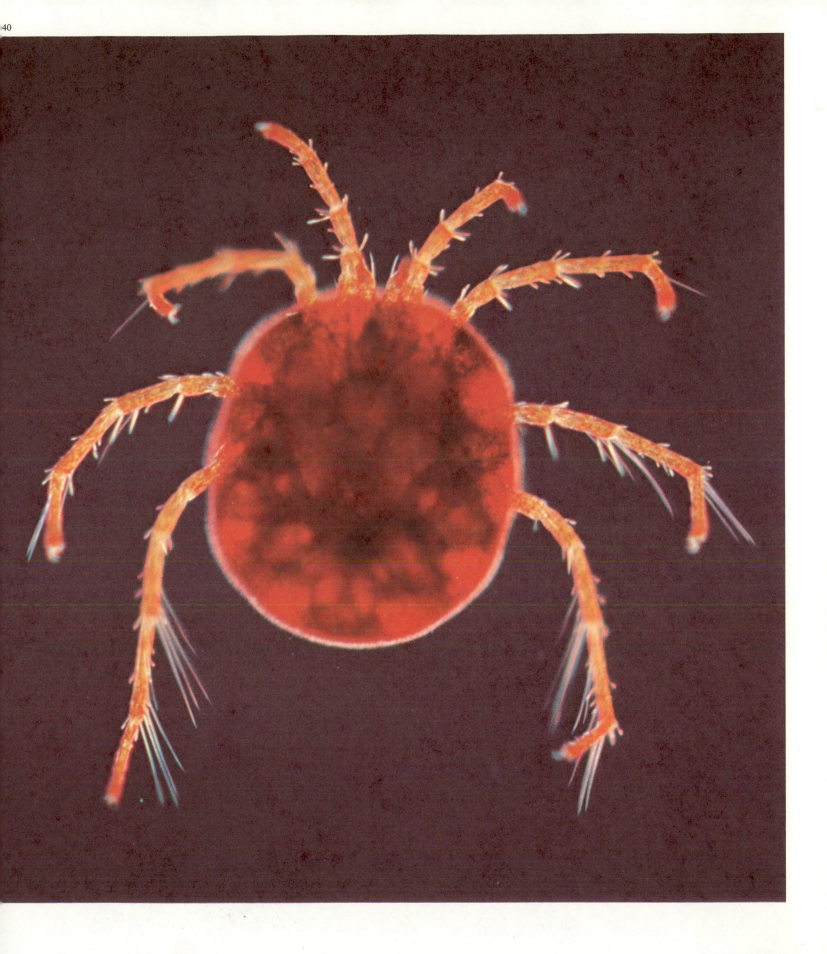

CHAPTER 4
SEED PLANTS

ACCORDING to modern science, life on Earth began about 3000 million years ago, in the warm waters of primeval oceans. Evolution has since produced the amazing diversity of creatures on Earth today. Early plant life consisted of simple organisms which lived and reproduced in water. Gradually plants invaded the land, but they remained dependent on external water for their successful reproduction.

About 350 million years ago, two inventions appeared which freed the land plants from their need for a watery or humid environment. One was the pollen grain. The other was the seed. The inventors were a group known as Gymnosperms. Today they are represented by conifer trees, cycads and the Japanese maidenhair tree, *Gingko biloba*.

The next evolutionary step took place about 130 million years ago. This was the appearance of the flowering plants, or Angiosperms, which dominate the plant kingdom today. Charles Darwin described the relatively sudden appearance of the flowering plants as 'an abominable mystery'. It was a remarkably successful event. There are about 250 000 species of flowering plants. They range in size from a tiny duckweed, *Wolffia arrhiza*, only about 1 millimetre across, to huge trees such as the coast redwood, *Sequoia sempervirens*, the tallest of which may reach 110 metres.

The success of the seed plants lies in their methods of reproduction and dispersal of their offspring. The male sex cells are produced within a tough, drought-resistant package, the pollen grain. This protects the sex cells during their initial journey to the female. After successful fertilisation, the offspring are packaged as seeds. These are dry structures, capable of surviving for long periods and travelling great distances without the need for water. Using these twin strategies, seed plants have colonised almost the entire land surface of the Earth.

The word 'flower' evokes a particular vision, of something coloured, possibly scented, beautiful – an adornment for our gardens and homes. Flowers corresponding to this vision in fact represent only a part of the wide spectrum of forms produced by evolution. Though largely unnoticed, flowers are produced in huge numbers by grasses, cereal crops and forest trees. The reason why some flowers are conspicuous and others are not is related to the method they use to achieve the successful union of male and female sex cells.

Inconspicuous flowers generally rely on the wind to carry their pollen grains for them. The wind cannot see, so there is no need to produce a showy flower. Coloured and scented flowers are designed to appeal to some living creature which will act as the bearer of the pollen. The most common pollinators are insects, but some plants use birds, bats, or even snails. They are drawn to the flower by sight or scent, and rewarded for their visit by gifts of nectar, or of some of the highly nutritious pollen itself. In specialised cases, the rewards are of a different nature. Many orchids, for example, pose as females of insect species. They are visited by hopeful males, who pollinate the flower while attempting to satisfy their own sexual desires.

The problems of living on dry land go beyond reproductive strategy. Life as a large land plant is altogether different from that as a small aquatic organism, let alone a single cell. Water is still needed for growth: it has to be extracted from the ground and conveyed to all parts of the plant body. The land plant therefore has to develop an extensive root system and an internal vascular network. In order to obtain energy for growth, the plant must expose green photosynthetic tissue to sunlight. For aquatic plants, this presents no mechanical problems – the water bears the weight of the plant tissues. On land, a strengthened stem has to be made to support the leaves. Exposed to the sun and the wind, the leaves themselves must be designed to reduce the loss of precious water to a minimum, while at the same time carrying on their photosynthetic function.

Throughout their evolution, plants have had to face predation from other living creatures. This hazard is particularly acute on land, where plants are prey to the attention of flying insects and birds, as well as animals.

Plants have solved these problems in a variety of ways. The 250 000 species of flowering plants could be said to represent 250 000 different solutions. It is only by the use of microscopy that the beauty and subtlety of some of them can be appreciated. The scanning electron microscope, in particular, has been a source of revelation to botanists. Many of the scanning electron micrographs in this chapter were produced by viewing the specimen in an intact frozen state. They show details of surface structure and function which could not have been revealed by any other method. Before the invention of the scanning electron microscope in the late 1940s, such pictures were unthinkable.

4.1 The first stage in the reproduction of a seed plant is the safe arrival of the pollen at the receptive female surface, called the stigma. In this scanning electron micrograph of a frozen flower of the turnip, *Brassica campestris*, a group of pollen grains is seen covering the stigmatic surface. Each pollen grain is egg-shaped, with a patterned surface and a furrow along its length. Two smooth projections from the female stigma are also visible. The pollen grain in the centre of the picture has produced a narrow pollen tube. This has penetrated the female surface. It will grow down into the flower until it reaches the vicinity of the egg cell. Inside the pollen tube, two sperms follow its growth. When the egg is reached, the tube bursts and releases the sperms, one of which will fertilise the egg. The whole process takes place without the need for external water.
SEM, false colour, ×350

ROOTS

Roots have two functions: they anchor the plant and extract water from the soil. Seeking water, they can grow to considerable depths. The roots of maize extend 1.5 metres down, but the roots of trees penetrate to 6 metres or more in light soils. The main root produces many branches, called laterals. Roots also produce fine hairs which serve to increase enormously the area of their absorbing surface. In a study of a four month old rye plant, it was estimated that the root system produced 14 billion root hairs and had a total surface area of over 600 square metres. This was 130 times greater than the surface area of the plant above ground.

In the tip of the root, special cells called statocytes are produced which can sense gravity and cause downward growth. They contain large and dense starch grains which act as gravity detectors. Later in life, these same cells are pushed outwards to the surface of the root tip, where they produce a slimy mucus to lubricate the passage of the root through the soil.

Water absorbed by the root hairs is transferred to the vascular system of the plant. In roots this system consists of a central cylinder of tissue containing *xylem* and *phloem* cells. The xylem cells carry water from the roots to the rest of the plant. The phloem cells bring supplies of sugar from the leaves to sustain the growth of the root tip.

4.2 Root hairs are short-lived, delicate structures. Roots grow from the tip, and new hairs are continually produced from a region just behind it. The process is clearly visible in this very low-magnification scanning electron micrograph of a frozen cabbage seedling. Near the root tip, the hairs are young and short; further

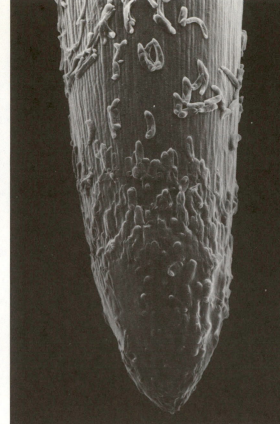

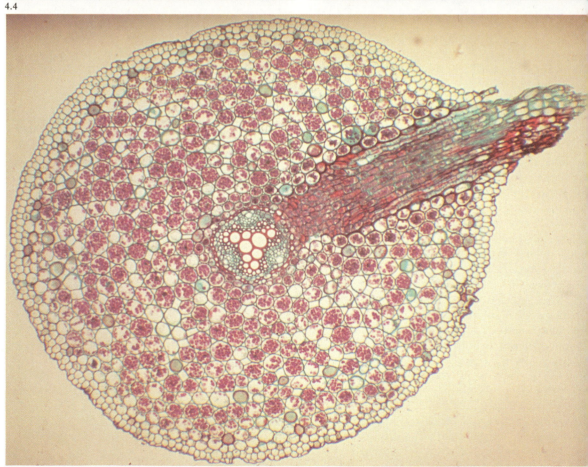

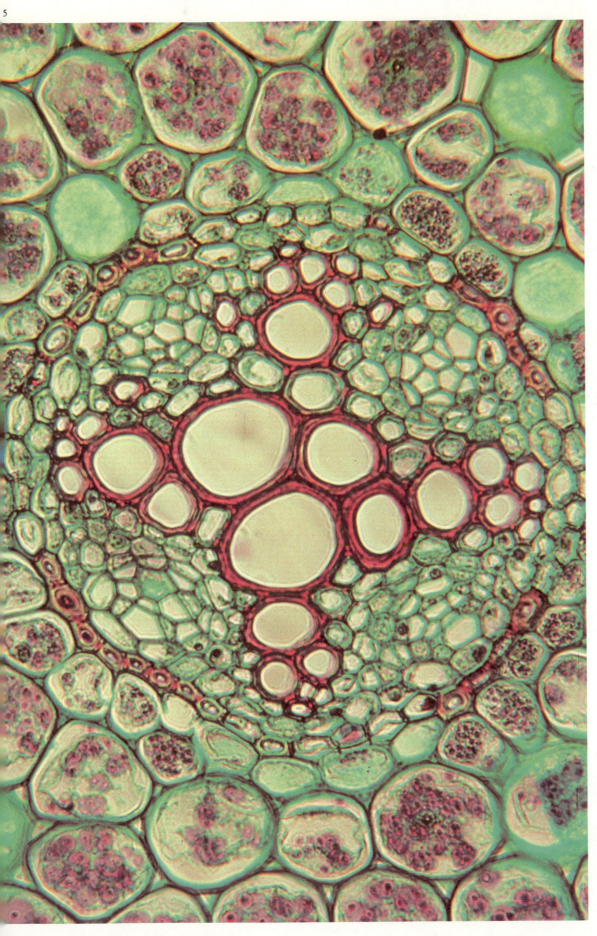

away they are longer. The picture also shows how the coat around the seed at the top right has split open to allow the root to emerge. Inside the split seed, the first leaves have yet to unfold.
SEM, ×12

4.3 The root tip has to push its way through the soil. To assist this operation, it continually produces a loose layer of slimy cells at its outer surface. In this picture of a laboratory-grown root of wheat, these cells can be seen embedded in the slime layer, which appears as a smooth covering on the frozen specimen. During growth in soil, the cells are rubbed off and die.
SEM, ×60

4.4 Laterals are produced deep within the main root from a special layer of cells surrounding the vascular tissue, called the pericycle. This light micrograph shows a developing lateral in the root of the buttercup, *Ranunculus acris*. As the young root lateral pushes outwards, it crushes and kills the cells of the root cortex which are in its way. The picture shows that many light microscope stains are not very specific in their action. The green dye has stained cell walls only, but the red dye has coloured starch grains (small particles within the cortical cells) as well as the thickened walls of the xylem cells in the centre of the root and other cells in the lateral itself.
LM, bright field illumination, stained section, ×22

4.5 In this close-up of the central region of a root similar to the one in Figure 4.4, the details of the vascular region, or stele, are revealed more clearly. In the centre, arranged in the form of a cross, are the large empty xylem cells, their walls stained red. Between the arms of the cross at the edge of the circular stele region are four groups of other cells, also red. These are the phloem. Beyond the stele lies the cortex of the root with its large cells containing red-stained starch particles.
LM, bright field illumination, stained section, ×870

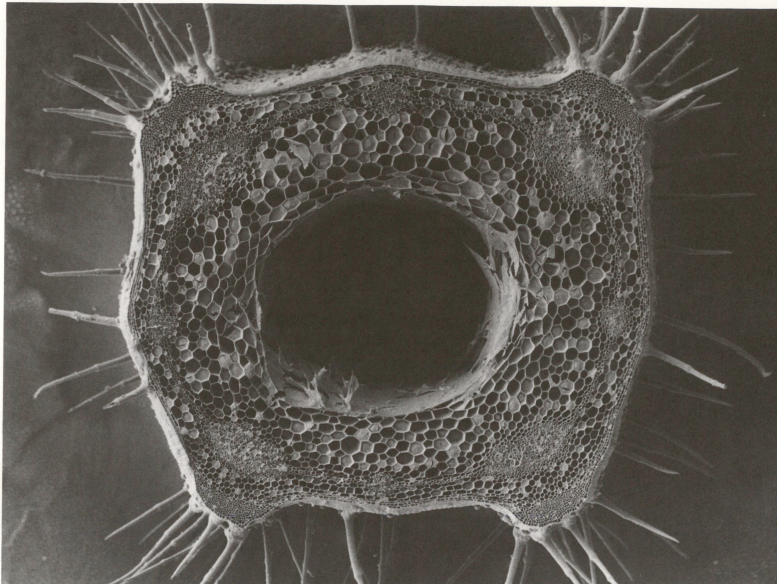

STEM & WOOD

The stem of a plant connects the leaves with the root system. Water and soil minerals travel up the stem in the cells of the xylem, and sugar solution flows down the stem in the phloem. The stem has an important additional function: it supports the entire weight of the aerial parts of the plant. This weight may be slight in the case of a small annual plant. A long-lived tree, however, may present a formidable problem. The most

massive living thing on Earth is a tree. The giant sequoia known as 'General Sherman' – a specimen of *Sequoiadendron giganteum* – is estimated to weigh over 2000 tonnes.

Stem growth is classified into two types. Primary growth occurs in annual plants. The stem tissues consist of vascular cells, together with unspecialised cortical cells. Strength is gained by the formation of groups of cells called collenchyma, which have very thick walls. Primary stems are often hollow, because a hollow tube resists bending better than a solid rod of the same weight.

Secondary growth occurs in long-lived perennial plants with

persistent stems, of which the most obvious examples are trees. Each season, the stem grows thicker by the action of a ring of soft tissue, the cambium, which is located under the bark. The cells produced towards the outside of the cambium become phloem; those towards the centre of the stem become xylem. The xylem cells develop massively thickened walls made of cellulose together with an inert filler called lignin. Once this thickening has occurred, the cells die, remaining as a system of extremely strong and resilient tubes for the passage of water. This is the tissue which we call wood.

4.6 The scanning electron micrograph of the cut stem of the white dead nettle, *Lamium album*, shows all the characteristics of primary growth. The hollow centre is produced by the collapse of thin-walled pith cells. Surrounding this core is a layer of unspecialised cortical cells – the cortical parenchyma. Eight 'vascular bundles' serve to conduct water and nutrients: one near each of the four corners, and one half way along each side of the stem. In the extreme corners of the stem – the most advantageous position mechanically – small collenchyma cells are visible. The spikes on the outside of the stem are hairs, designed to discourage crawling insects from climbing the plant.
SEM, ×30

4.7 Trees grow in annual cycles, and this gives rise to the familiar growth rings visible on sawn logs. In this radial section of wood from the conifer *Tsuga canadensis*, the hemlock tree, the cycle is shown in terms of the width of the cells comprising each year's growth. The four bright red bands running diagonally from top right to bottom left correspond to four years of growth in the tree's life. Early in the season, the cells are widely spaced, but they bunch closer together as autumn approaches. Thus spring of one year is at the very top left of the picture; the bright red lines are widely spaced until, as autumn comes, they bunch together to form the thick red band; then, in the following spring, the pattern is repeated. The darker red bands running at right angles to the annual rings are 'rays', a system of horizontal xylem cells. The bright colouring of the micrograph results from the interaction of polarised light with the ordered cell wall structure of the xylem.
LM, polarised light, ×660

4.8 The structure of hardwoods is more complicated than that of the softwoods produced by conifers. There are more different types of cells and, in particular, hardwoods have very large water-conducting elements called vessels. This cross-section of hardwood from the stem of the sugar maple tree, *Acer saccharum*, was produced by the same polarised light technique as the previous picture. The large blue cells are the vessels. The red bands are cells of the horizontal xylem system, the rays, which distribute water to the cambium and also act as storage centres for starch and lipids. There are no annual growth rings visible in this section.
LM, polarised light, magnification unknown

Leaves are the site of
photosynthesis – the conversion of
atmospheric carbon dioxide into
sugar. This chemical process
provides the plant with energy
and, as a waste product, produces
the oxygen we breathe. The
chemistry takes place inside leaf
cells, in organelles called
chloroplasts, which are described
on pages 116–117 in the chapter
on the cell.

The raw materials of
photosynthesis are sunlight, water
and air. The first two present little
problem. The typical leaf is a flat
blade designed to catch the light,
and its water supply comes via the
stem from the roots. The difficult
supply problem concerns the air.
A leaf exposes a large surface area
to the atmosphere and, because of
this, it has to be covered with a
waterproof layer called the cuticle.
Without it, the plant would lose
water too rapidly, and die. But the
cuticle also stops air getting into
the leaf, so it has to be perforated
with a series of pores. These
pores, or *stomata*, are held open
during the day, and are closed at
night in order to conserve water.

The surface of a leaf is rarely a
flat landscape when seen through a
microscope. A wide variety of
specialised features comes into
view. These outgrowths are
collectively known as trichomes,
and scanning electron microscopy
reveals their diversity particularly
well, as Figures 4.14–4.18 show.

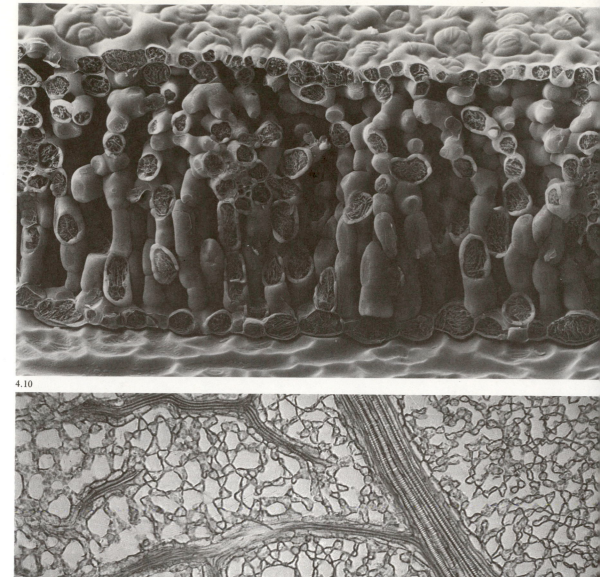

4.10

4.9 A leaf is built like a sandwich. To
see inside it with a scanning electron
microscope, the leaf must first be
frozen and broken open. This picture
is of a leaf of the turnip, *Brassica
campestris*, prepared in this way. The
single horizontal lines of cells towards
the top and bottom of the picture form
the skin, or epidermis, of the leaf,
covered with the cuticle layer. Special
cells in the epidermis control the size
of the stomatal pores, most of which
occur on the underside of the leaf,
which is at the top of this picture. The
loosely packed cells in the interior of
the leaf are called mesophyll cells, and
they are the site of most of the
photosynthesis. Some are intact,
others are broken open.
SEM, ×250

4.10 An entirely different view of a
leaf is given by the light microscope.
This section of the leaf of a privet,
Ligustrum vulgare, is dominated by the
leaf's vascular system, which is seen as
a branching network. The mesophyll
cells, loosely packed together between
the vascular strands, are small and
irregularly shaped; the large clear areas
amongst them are air spaces.
LM, bright field illumination, ×180

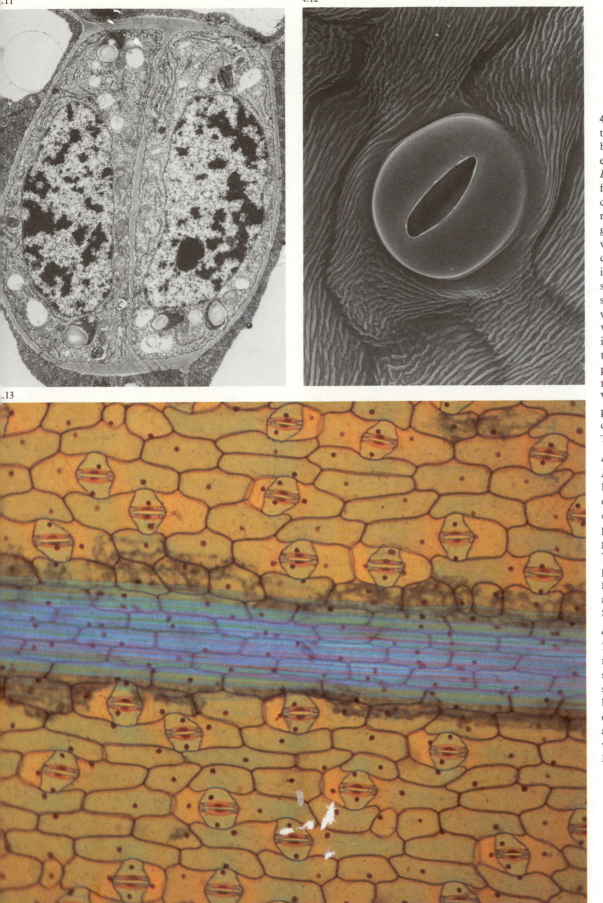

4.11 To see details within cells, a transmission electron microscope must be used. This pair of cells in the epidermis of a leaf of the garden pea, *Pisum sativum*, is at an early stage of forming a stomatal pore. Each cell contains a large centrally placed nucleus, oval in outline and containing genetic material stained black. Also visible are a few starch-containing chloroplasts – the five dark grey bodies in the cytoplasm, four of which are seen to have pale grey contents (the starch). The wall between the two cells will soon split down the middle. This will open an airway into the leaf's interior, which is effectively behind the page as it is viewed. The size of the pore will depend on the activity of the two cells, which are called guard cells. When they swell up, the pore will be pushed open; when they shrink, it will close.
TEM, stained section, ×5000

4.12 This stomatal pore on a sepal of *Primula malacoides* at first appears to be open. In fact, it is closed. The opening is a permanent gap in the cuticle; beneath it can be seen, tightly pressed together, the walls of the guard cells which control the pore. The ridges on the sepal surface are a peculiarity of this plant; the deeper furrows reveal the outline of the adjacent epidermal cells.
SEM, ×750

4.13 This surface section of a leaf of *Tradescantia* under the light microscope shows the distribution of stomata. The paired guard cells are stained red-brown; the small dark brown particles elsewhere are nuclei of epidermal cells. The broad blue band across the picture corresponds to a vascular strand, or leaf vein.
LM, ×445

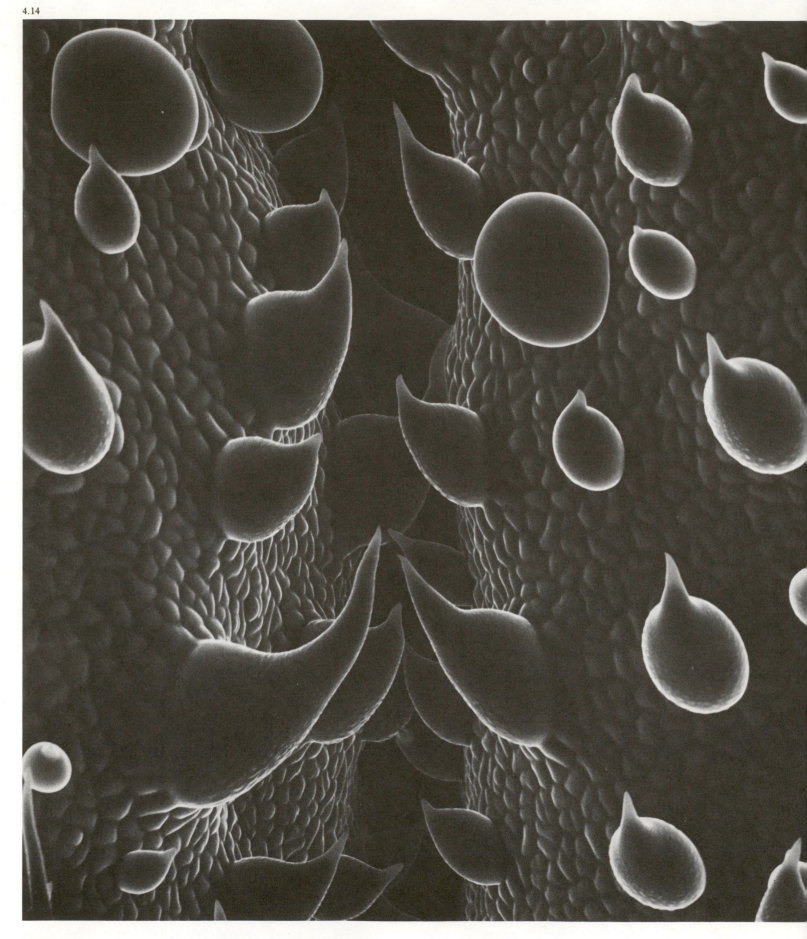

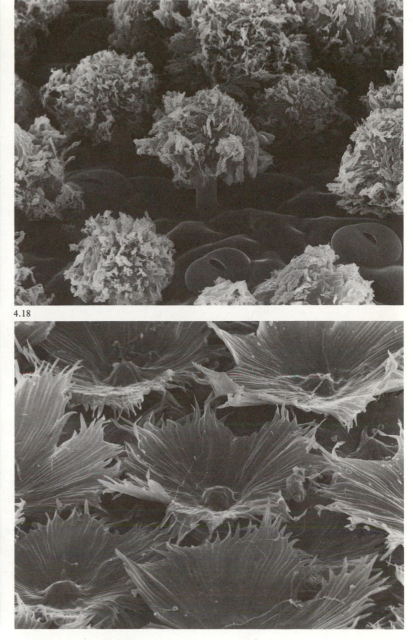

4.14 Flowering plants produce a wide range of substances designed to reduce their palatability to animals. The trichomes on this leaf of the marijuana plant, *Cannabis sativa*, are sites for the formation of two of these 'secondary products'. The pointed outgrowths are cystoliths, which contain tough crystals of the mineral calcium carbonate. The rounded structures towards the top of the picture are glandular trichomes containing tetrahydrocannabinol resin. It seems ironic that evolution has ensured the survival of this species by means of a poison which has proved highly attractive to human beings.
SEM, ×850

4.15 These branched trichomes completely obscure the surface of the leaf which produces them. Designed to reflect sunlight and keep the leaf cool, they belong to the plant *Verbascum pulverulentum*, popularly known as hoary mullein. The thick, hairy covering of trichomes was scraped from the leaves by neolithic flint miners at Grimes Graves in Norfolk. They are believed to have rolled it like cotton to use as wicks in oil lamps.
SEM, ×105

4.16 *Primula malacoides* is a popular pot plant which looks as though it has been dusted with flour. The reason is the production of wax from stalked trichomes, seen here on the surface of a sepal. The wax, called farina, insulates the plant and prevents condensation which would otherwise block the stomatal pores between the trichomes.
SEM, ×335

4.17 This branched trichome is not on the surface of a leaf, but inside it. The leaves of the water lily, *Nymphaea alba*, are permeated with wide passageways which supply air to the submerged roots via the stem. The internal surfaces of these airways are covered with hard trichomes encrusted with crystals of poisonous calcium oxalate. Their function is to deter inquisitive burrowing insects.
SEM, ×120

4.18 Plants have to conserve water, particularly in winter when it is difficult to obtain from cold or frozen ground. The Japanese evergreen shrub *Eleagnus pungens* employs these umbrella-like trichomes for this purpose. Found on the underside of the leaves, the trichomes overlap and completely protect the leaf from the effect of drying winds.
SEM, ×80

ATTACK & DEFENCE

Plants have an ambivalent relationship with the animal kingdom. They often depend on insects or other animals to achieve pollination of their flowers and dispersal of their seeds. But they also represent nothing more than a free lunch to hordes of bugs, aphids and caterpillars, not to mention more formidable adversaries in the shape of herbivorous mammals. Plants guard against such predation by synthesising poisons, or by covering themselves with spines, sharp bristles, or sticky glues.

Some of these defence mechanisms are astonishingly subtle. All species of potato, for instance, respond to wounding by the rapid production of a chemical which inhibits the enzymes used in an insect's digestive system; in effect, they make themselves indigestible.

Such chemical defences do not always work as intended. Plants of the milkweed family contain a strong cardiac poison, fatal to vertebrates. Insects, which are unaffected by it, have learned to gain immunity from their own vertebrate predators by eating the tissues of the plants.

Plants are not always passive, defensive creatures. Some inhabitants of boglands are able to obtain the nitrogen and minerals which the soil lacks by capturing and digesting insects. There are about 500 species of these carnivorous plants, and they have developed a variety of methods for immobilising their meal. One example is the sundew.

4.19 The upper surface of a sundew leaf is covered with stalked trichomes, each bearing at its tip a gland which secretes a strong adhesive. This scanning electron micrograph of a frozen leaf of the Cape sundew, *Drosera capensis*, shows the glands in

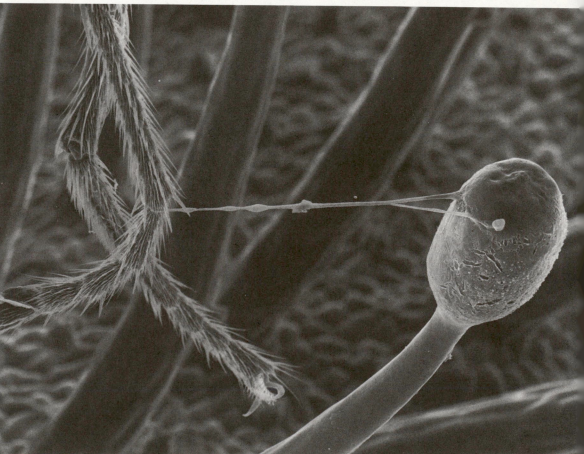

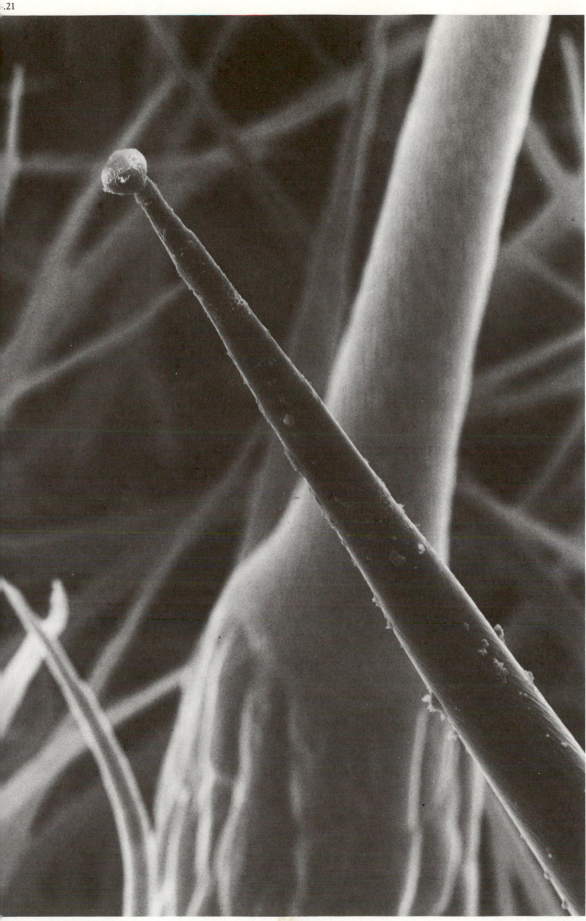

action. A small fly of the Psilidae family has wandered onto the leaf, and is stuck fast. The more the fly struggles, the more glands it touches, until eventually it becomes immobilised.
SEM, ×30

4.20 This detail of the same specimen as in Figure 4.19 shows that the initial contact between plant and insect may be very tenuous. Two of the fly's legs appear in the picture, one showing the pair of tiny claws which it uses for gripping rough surfaces. A very thin strand of adhesive attaches the other leg to the tip of a trichome. At this stage, the fly may escape if it is strong enough. But if it doesn't, its struggles will set up electrical signals in the trichome's stalk. Over a period of perhaps 30 minutes, these signals will cause the stalk to bend inwards towards the fly. Once in contact with the fly's body, the gland secrets a digestive juice and absorbs the nutrients which this releases from its prey.
SEM, ×125

4.21 The leaves and stem of the stinging nettle, *Urtica dioica*, are covered in hairs. The majority of them are of simple construction, but a few are modified into stings. Stings occur in groups of two or three, mostly along leaf veins. This picture shows both the base of a sting (the thick structure in the background) and the tip of another sting in the foreground. The wall of the sting is brittle and contains silica. At its tip, a small spherical cover of a glass-like material seals the end. The slightest touch to this seal causes it to break, uncovering a sharp hollow needle finer than a hypodermic. The injection which results is a cocktail of two chemicals, acetylcholine and histamine. The nettle's sting is an effective defence against grazing animals but largely useless against insects.
SEM, ×250

4.22

4.23

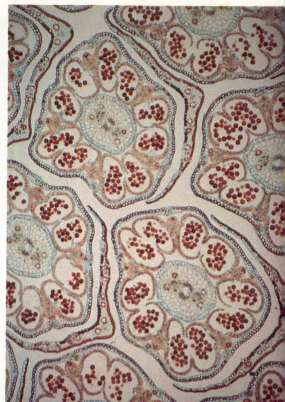

The bewildering diversity of form, colour and scent of flowers has one purpose – the reproduction of the plant. The idealised flower consists of a series of different structures arranged in concentric layers. On the outside there is a layer of protective sepals, then a ring of petals, then a ring of male organs called stamens. Each stamen consists of a stalk with an anther at its tip. The anther is the site of pollen production. In the centre of the flower is the female tissue, the carpel. This contains the egg cells, and it also develops a receptive surface called the stigma. In many flowers the stigmatic surface is produced at the end of a stalk growing from the top of the carpel. In any particular flower, each of these layers may be modified, or absent.

Flowers may be produced singly, or in complex groups comprising thousands of individuals, as in the daisy family. Some are microscopically small; on the other hand, a species of *Rafflesia* produces a bloom nearly 1 metre in diameter and weighing 7 kilograms. Nor are all flowers delicate and innocent; the African water lily *Nymphaea citrina* systematically drowns its insect visitors to wash pollen from their corpses onto its female parts.

4.22 A flower begins its life as a series of small bumps on the side of a shoot tip. Its final form is not generated until after considerable growth. This small bud of the snapdragon, *Antirrhinum majus*, appears almost symmetrical; the flower it produces is highly asymmetric. The sepals have been dissected away so that the outer, five-lobed layer in the picture represents the petals. Inside them, the tips of four developing anthers can be seen. In the centre of the bud, the cleft surface of the stigma has formed.
SEM, ×80

4.24

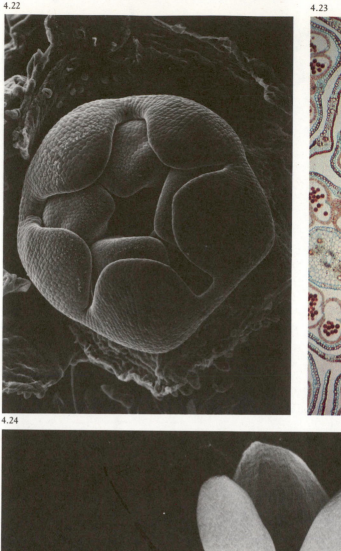

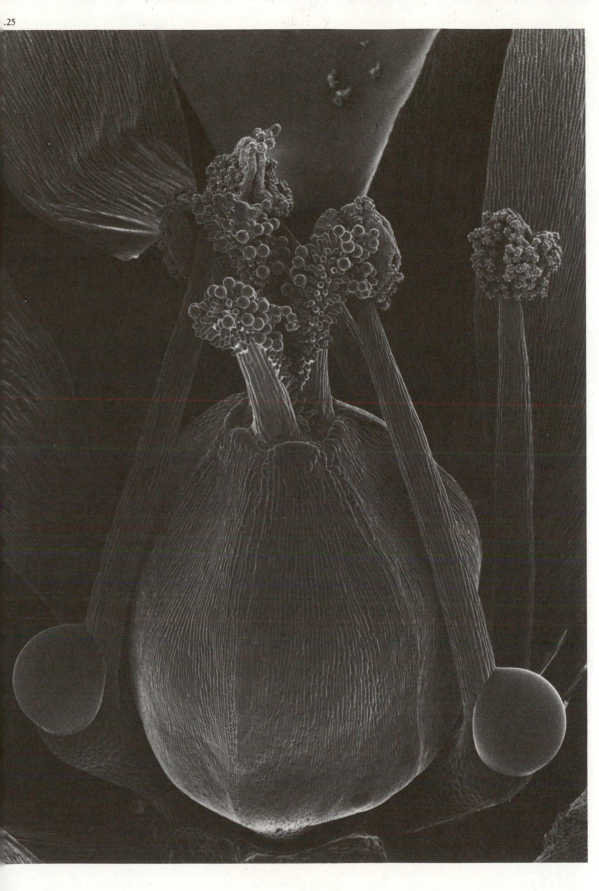

4.23 Each 'flower' of a daisy such as *Cosmos bipinnatus* is a mass of thousands of tiny floral units or florets. Those at the edge – the 'petals' – are sterile. This light micrograph shows a section through a group of fertile florets in the centre of the *Cosmos* inflorescence. Each floret contains five paired anthers, recognisable by the mass of brown-stained pollen grains within them. They surround the carpel tissue, stained pale blue; the developing ovules within the carpel are the small orange circles.
LM, bright field illumination, ×50

4.24 The flower of the chickweed, *Stellaria media*, conforms closely to the idealised form. This low-magnification scanning electron micrograph shows the five sepals on the outside, with five deeply lobed petals inside them. The three granular objects are anthers, covered in pollen grains. In the centre of the flower is the carpel, with three rough stigmatic surfaces visible.
SEM, ×15

4.25 This picture is of the same species as in Figure 4.24. The view is from the side, and was obtained by dissecting away part of the flower and tilting it in the microscope. It shows the bulbous carpel, with the stigmas emerging from it on the end of three short stalks, or 'styles'. Two of the long-stalked anthers have moved to make contact with the stigmas; the one on the right of the picture has remained separate and clearly visible, its tip covered with pollen. The smooth round objects at the base of two of the anther stalks are drops of nectar. This is a sugary substance offered to insects as a reward for visiting the flower and pollinating it. Curiously, a chickweed is quite capable of pollinating itself and often does so in the wild. The round pollen grains visible on the stigmatic surfaces of this flower show that it has been pollinated, probably by the two anthers in contact with the stigmas.
SEM, ×40

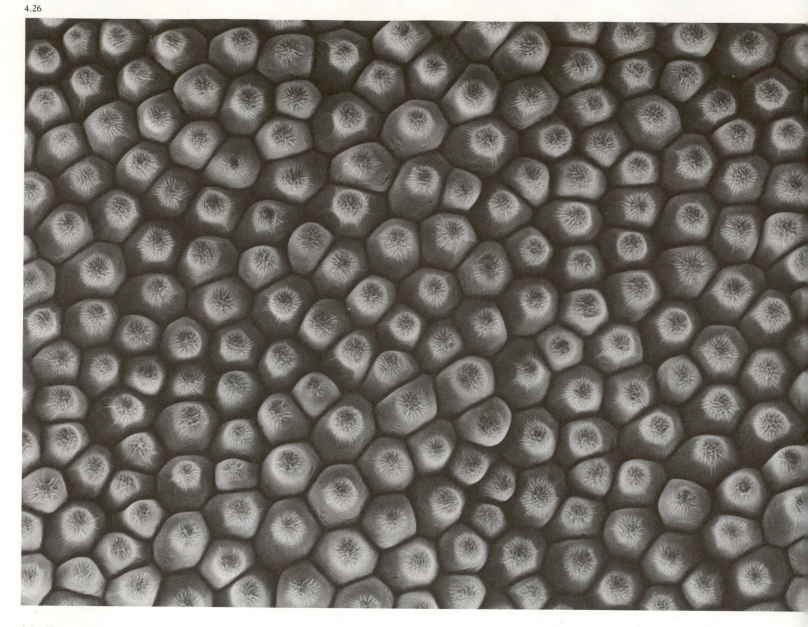

4.26 The petals of most flowers are in fact stamens which have switched roles during evolution from sexual organ to advertising agent. Although they appear smooth to the eye, they often reveal fine sculptured detail when viewed with a microscope. This rose petal surface consists of closely packed cells, each one of which is topped with fine ridges. The reason for this is unknown, but may be related to the reflection of light from the petal. SEM, ×365

4.27 The stigma is the first point of contact between the pollen and the female organs which it is to fertilise. The stigmatic surface is usually moist and sticky, and often feathery or brush-like; this increases the area available for pollen to land on, and so improves the chances of a successful

union. In this scanning electron micrograph of an *Hibiscus* flower, five separate stigmas are visible. Not every pollen grain which lands on a stigma is accepted by the female. Cross-pollination between different plants of the same species is the general rule in plant reproduction. The stigma will inhibit the growth of pollen which is not of the correct species. In many cases, it will also reject pollen from other flowers of the same plant. SEM, ×20

4.28 Flowers are not all hermaphrodite. Some are exclusively male, others female. In such cases the tissues corresponding to the other sex are absent from the flower. This group of florets from the common daisy, *Bellis perennis*, are all female. The picture shows the developing stigmatic

surfaces at the tips of a group of carpels. There are no stamens present. Interestingly, most of the flowers are producing five stigmas, but one has developed abnormally and has six. SEM, ×35

4.29 The anther first develops as a closed chamber divided into four spaces. As the pollen grains develop, the wall between pairs of spaces is broken down, and at maturity the anther consists of two closed pollen sacs. This mature anther is from shepherd's purse, *Capsella bursa-pastoris*. The coarsely textured surfaces are the outside of the anther wall. The finer surface is the inside of the same wall. It is visible because the anther has split open to allow the pollen to escape – a process called dehiscence. A dozen or so pollen grains remain inside

the opened pollen sac. SEM, ×220

4.30 This picture is a close-up of one of the anthers of the chickweed specimen in Figure 4.24. In chickweed, the pollen sacs open·very wide indeed, effectively turning themselves inside out. This results in the appearance shown here, of a mass of pollen grains lightly adhering to the inside wall of the sac. In wind-pollinated flowers, the grains are wafted away on air currents. Chickweed pollen simply waits. If an insect happens to pass, then grains may be brushed onto its body and effect a distant act of cross-pollination If no insect appears, the anther stalk eventually bends inwards so that the pollen will fertilise its own flower. SEM, ×150

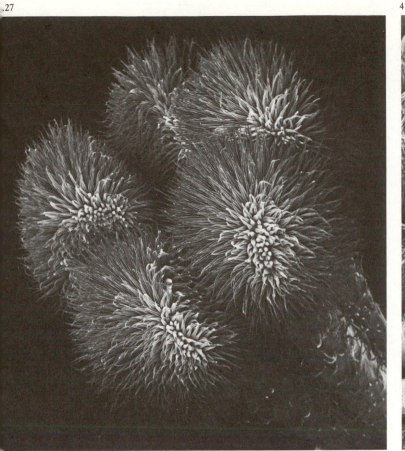

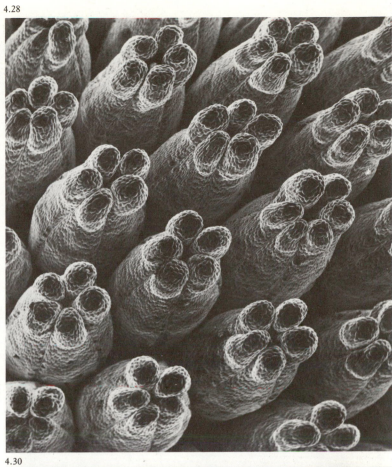

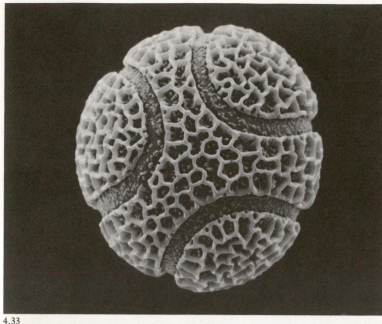

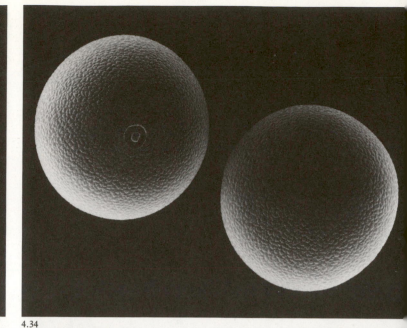

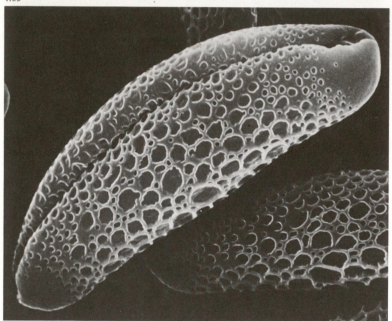

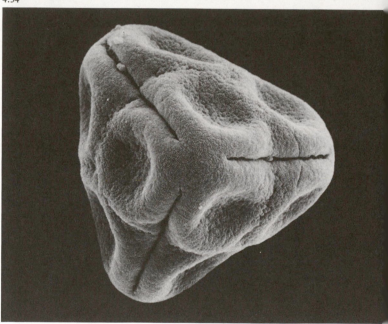

POLLINATION

All pollen looks much the same to the unaided eye – like fine yellow dust. Microscopy reveals that pollen grains differ widely in size, shape and surface texture. Those of the alpine forget-me-not, *Myosotis alpestris*, are only 3 micrometres across, whereas cucumber pollen (*Cucurbita pepo*)

has a diameter of 200 micrometres. About half the flowering plants produce ellipsoidal pollen grains, but spheres, polyhedra and long thin rods are also encountered. The surface varies from almost smooth to highly textured; small spikes, ridges and furrows appear in intricate patterns. Smooth pollen grains are found particularly in wind-pollinated flowers; they separate from each other easily in preparation for flight. Textured grains are designed to adhere together and to the hairs of pollinating insects.

The surface pattern on a pollen grain is in some cases characteristic enough to allow the species to be identified. The wall contains a very inert polymer called sporopollenin, as a result of which the pattern may be preserved for hundreds or thousands of years. This is true of pollen preserved in peat bogs, and is of archaeological value.

The task of the pollen grain is to carry the male sex cells to the female stigma. If the partners are compatible, the pollen grain germinates to produce a thin tube

– the pollen tube. This grows through the stigmatic surface and down to the ovule, where it bursts releasing two sperms.

Compatibility depends on the recognition of proteins released by the pollen. These can be highly allergenic to humans. Hay fever results from breathing in pollen grains of wind-pollinated flowers, especially grasses. The amount of pollen produced by such flowers can be enormous; a single catkin of the hazel, for example, produces about four million pollen grains.

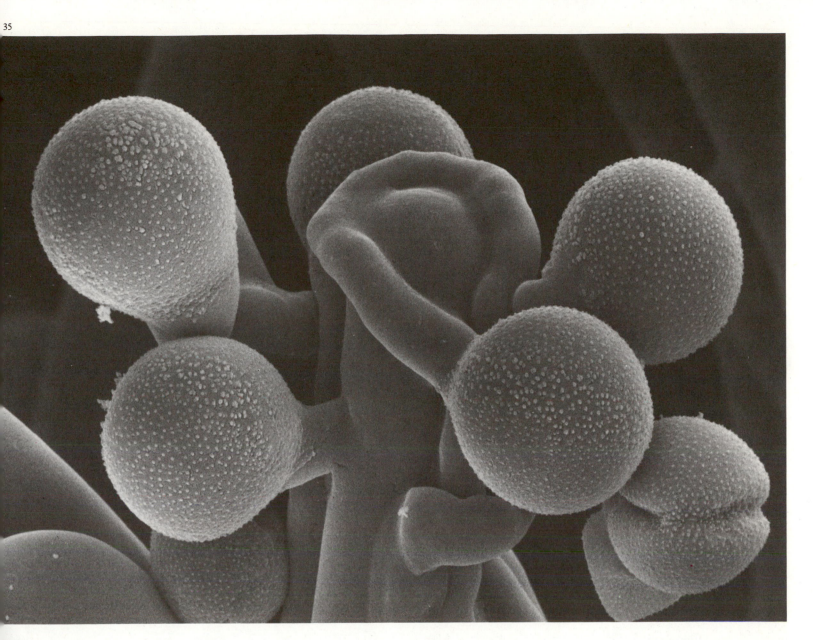

4.31 Pollen grains are a favourite specimen for scanning electron microscopy. Not only beautiful, they are often tough enough to need no chemical fixation or freezing. The single grain in this picture is from the passion flower, *Passiflora caerulea*. Passion flowers provide copious nectar, and are insect-pollinated. A newly opened passion flower has its anthers fully exposed, but no female parts. Over the course of a few days, it unloads its pollen onto passing insects, then the anthers shrivel. Their place is taken by newly developed stigmas on the ends of elongated styles. This separation of the sexes guarantees cross-pollination. The pollen tube emerges from one of the three curved furrows visible here.
SEM, ×1275

4.32 This spherical pollen is typical of many grass species, which are wind-pollinated. It is from the cocksfoot grass, *Dactylis glomerata*, a plant responsible for much early summer hay fever in Europe and North America (where it is known as orchard grass). The allergenic proteins are localised in the germination pore of the pollen wall. The grain on the right has its pore visible.
SEM, ×1065

4.33 The pollen of *Billbergia nutans*, a Brazilian member of the pineapple family, has a deep furrow along its length. The pollen tube is produced at one end of this furrow.
SEM, ×1475

4.34 This particle is not a single pollen grain, but four grains arranged in the form of a tetrahedron. All pollen grains initially develop in groups of four, called tetrads. In most plants subsequent growth results in their separation. In some species, such as this heather, *Erica carnea*, separation does not occur. Three of the four members of the tetrad can be seen. Each has three shallow depressions on its surface, and is bounded by a deep furrow through which the pollen tube or tubes will emerge.
SEM, ×1900

4.35 This view of germinating pollen grains of the opium poppy, *Papaver somniferum*, was obtained by freezing the flower before putting it into the scanning electron microscope. Eight pollen grains appear in the picture, clustered round a single finger-like projection, part of the stigmatic surface. Several pollen tubes are visible, most notably the one which encircles the top of the stigma before eventually growing downwards towards the base of the flower. The pollen grain at bottom right has not germinated; it retains the surface furrows which are characteristic of the species. Once germination occurs, the grain swells and the furrows disappear. At least five of the pollen grains in the picture have germinated, and this will result in the formation of a corresponding number of seeds. These develop from the ovules within the carpel, and as this occurs, the carpel tissue itself expands to produce the familiar seed capsule of the opium poppy. A single capsule may contain 2000 seeds.
SEM, ×1080

EMBRYO

Pollination and the growth of the pollen tube result in the release of two sperms into the embryo sac, the part of the ovule which contains the egg. One sperm fertilises the egg itself. This produces a cell called a zygote, which will eventually grow into the embryo. The other sperm fuses with two nuclei to form a special tissue called endosperm. The endosperm grows quickly, and serves as a source of nutrition for the young embryo.

These events all occur within an ovule. A flower may contain many ovules, each capable of producing one seed if it is fertilised. Apart from the embryo sac, the ovule consists of a nutritional tissue called the nucellus, together with a covering, or integument.

The development of plant embryos follows a variety of patterns, depending upon the species. In broad outline, the zygote divides and initially produces a short filament of cells, known as the suspensor. One end of the suspensor is anchored to the embryo sac, while the other end grows into the endosperm tissue. At this free end, the terminal cell divides repeatedly to produce a compact mass of small cells which are initially all the same in appearance. Once this globular mass reaches a certain number of cells, usually a few hundred, specialisation of the tissue begins to occur. The shoot tip, root tip, vascular system and first elementary leaves are formed. After this, the embryo grows rapidly, laying down storage materials such as starch and protein to sustain it during the initial stages of its eventual germination. The food supply for the growing embryo comes from the tissues around it, but also from the parent plant through the stalk

connecting the ovule to the carpel wall.

The final stage in embryo development involves loss of water. The embryo shrinks in size and the connections between it and its parent are severed. The ovule develops a hard protective covering from the integument layers. Finally the embryo is dry, full of food reserves, and protected by a tough coat. It has become a seed.

4.36 Conifers belong to the group of plants called Gymnosperms, which means 'naked seeds'. This refers to the fact that the ovules, and hence the seeds, are carried on the outside of the female flower, not enclosed in a carpel. In this light micrograph of a female pine cone, the ovules appear as egg-shaped objects around the core of the cone. They are accessible to wind-borne pollen which falls between the covering of scales.
LM, magnification unknown

4.37 The other group of seed plants is the Angiosperms, meaning 'enclosed seeds'. To obtain this view of the ovules of the opium poppy, *Papaver somniferum*, the developing seed capsule had to be cut open. The picture shows the ovules attached to the mother plant by means of a short stalk, the funiculus. Each ovule will eventually become a seed.
SEM, ×45

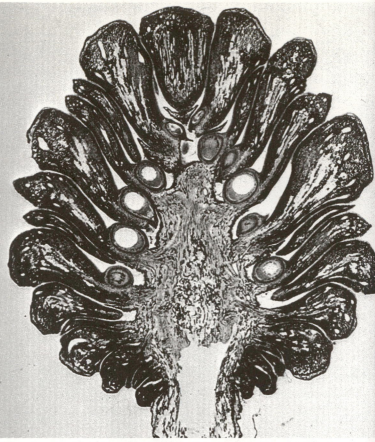

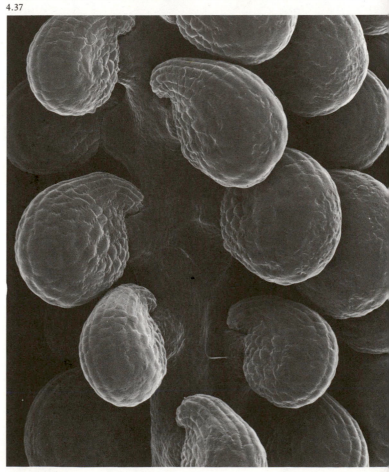

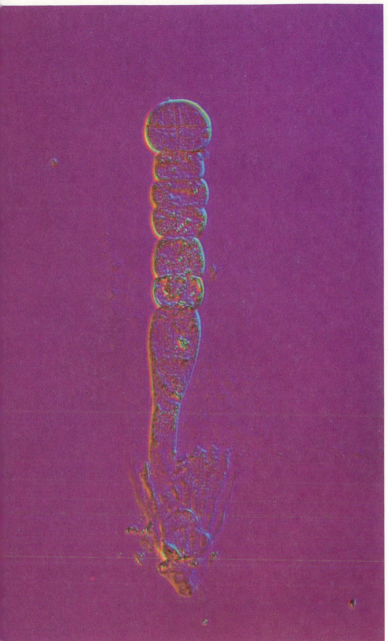

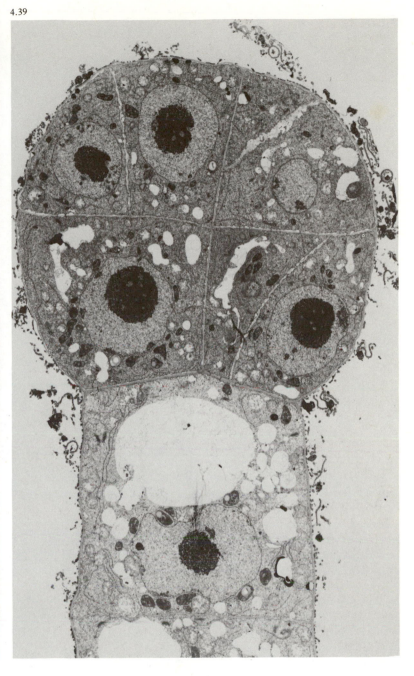

4.38 This embryo of the turnip, *Brassica campestris*, has been dissected out of the ovule two days after pollination took place. It consists of ten cells: six in the elongated suspensor, and four in the small globular tip at the top. The material at the bottom of the suspensor is remnants of the embryo sac. The cells of the globular tip will go on to produce the embryonic shoot and leaves, while the root will be formed from the suspensor cell nearest to the globular tip. The colours result from use of polarised light in the Nomarski Differential Interference Contrast microscope.
LM, Nomarski DIC, ×385

4.39 This transmission electron micrograph shows the globular tip and the first suspensor cell of an embryo like the one in Figure 4.38, but after one day's additional growth. The globular tip now contains sixteen cells, although this section, which is just 70 nanometres thick, only includes seven of them. The large grey bodies with black centres which can be seen in most of the cells are their nuclei. The large white areas in the suspensor are water-filled spaces called vacuoles. The material adhering to the outside of the embryo is fragments of endosperm tissue.
TEM, stained section, ×2600

4.40

Seeds have two purposes. They enable plants to survive through cold or drought – without seeds there would be no annual plants in temperate or arctic regions. And they make it possible for plants to colonise new habitats.

The dispersal of seeds involves a variety of mechanisms. Some seeds are simply very small and light, and are carried by air currents. The winter-flowering succulent *Kalanchoe blossfeldiana* has seeds which weigh only 1/100 000th of a gram. Others develop wings or propellers and can fly, sometimes for great distances. Some attach themselves to passing animals by stealth, but others are presented within colourful, tasty fruits which animals, including humans, collect and carry away.

Because of their hard coat and dry, dormant state, seeds can survive for long periods in the soil. Common weeds such as field poppies produce seeds which may live for 30 years or more until chance brings them to the surface and they grow. The record for seed longevity is held by the Indian lotus, *Nelumbo nucifera*, seeds of which have germinated after 1000 years of lying on a lake bed.

4.40 Small seeds, such as this one from the snapdragon, *Antirrhinum majus*, are often very rough in surface texture. The ridges and craters are produced by outgrowths of the integument of the ovule. Their purpose is to trap small particles of soil, so that when the seed falls to the ground it is anchored and can germinate in security without being blown away.
SEM, ×150

4.41 Goosegrass, *Galium aparine*, disperses its seeds by means of tiny hooks which cover the outside of the fruit. The hooks here are in the process of formation. The two young fruits are still firmly attached to their common stalk. At maturity, this attachment will become tenuous and any passing furry animal, or clothed human, may be enlisted to disperse the fruit. This technique of seed dispersal inspired the invention of Velcro (see page 179).
SEM, ×140

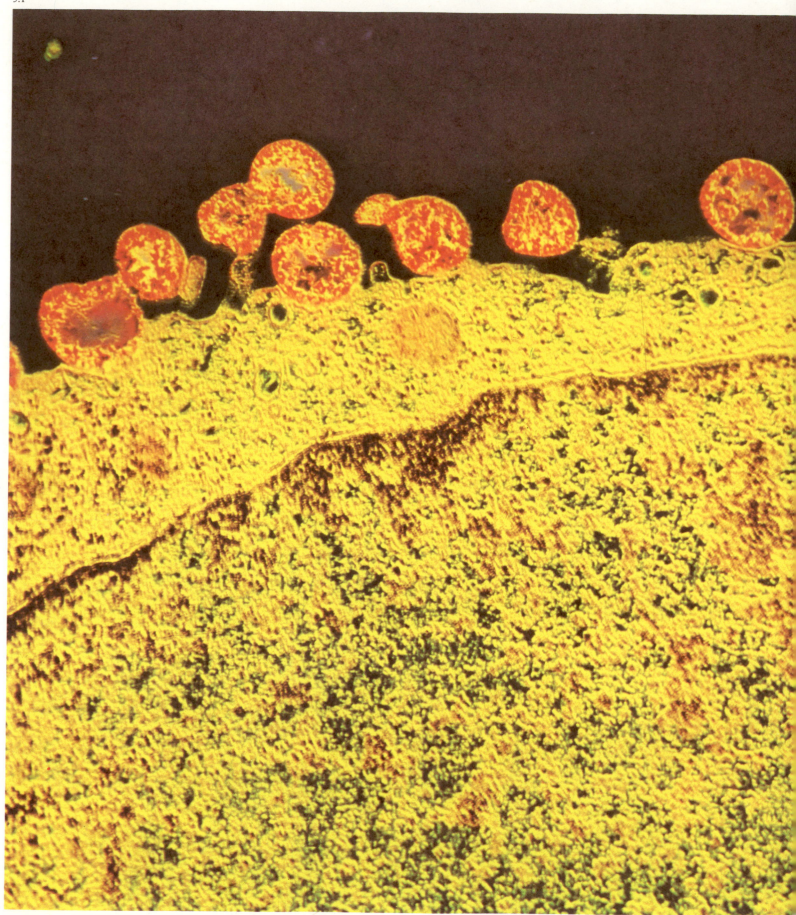

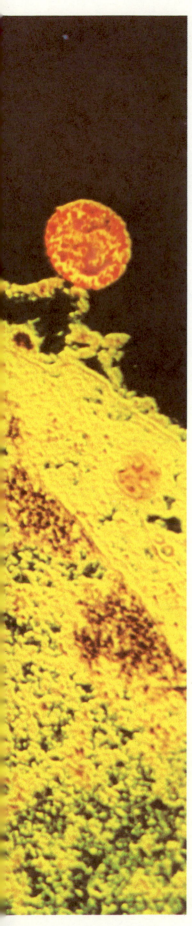

CHAPTER 5

MICRO-ORGANISMS

MICRO-ORGANISM is a popular term that is commonly applied to a wide biological spectrum. In fact, some micro-organisms are not alive, and others produce forms which are not microscopic. In this chapter, the term is used to include viruses, mycoplasmas, bacteria, algae and fungi. Protozoa have been covered already in the chapter on animals.

Micro-organisms were first seen by Antoni van Leeuwenhoek, the 17th century Dutch microscopist. He described various bacteria and protozoa, grouping them all together as 'animalcules'. Later developments have shown us that such a simple classification is not tenable. We can now see not just the external form, but details of the internal structure of micro-organisms, and their great variety is apparent. Some are indeed like 'little animals' in structure and behaviour, but some are plant-like, and in many, such as bacteria, the concept of plant or animal is irrelevant. The viruses are not living organisms at all.

Without microscopy the identification, description and study of micro-organisms would be impossible. The nature of viruses, for instance, only began to emerge after electron microscopy was first used to see them, in the late 1930s. Given a sample of micro-organisms, whether from water, soil, or diseased tissue, a biologist's first action is to look at it using a microscope.

Most micro-organisms are very small indeed. The tiniest are filamentous viruses, discovered in 1963. Consisting of a DNA molecule associated with protein, they are 5.5 nanometres (5.5×10^{-9} metres) thick, and about 1000 nanometres long. Other viruses form particles 20–100 nanometres across. Viruses are too small to be resolved in a light microscope. The smallest living cells, on the other hand, are just visible by optical microscopy, but need the electron microscope's high resolution to be seen in detail. Mycoplasmas and the smallest bacteria are about 0.3 micrometres (0.3×10^{-6} metres) across. Larger cellular micro-organisms such as yeasts and single-celled algae have dimensions of 1–100 micrometres. Some algae and fungi produce long filamentous cells which are just visible to the naked eye, and these may form large complexes. Giant kelps, for example, have fronds up to 150 metres long and are amongst the largest living things.

Scientific classification is based on biology and structure, not size. The most elementary micro-organisms are viruses, which rely on living cells for their reproduction. The smallest cells are also mostly parasitic, although some mycoplasmas and small bacteria can be cultured *in vitro* in complex nutrient media. Larger bacteria are usually free-living and adaptable; different species can survive in habitats as diverse as spring water or mineral oil.

Such simple cells are prokaryotic: their genetic information – the DNA – exists as a naked molecule within the cell. All other cells, including those of algae and fungi, are eukaryotic: their DNA is enclosed within a body called a nucleus. This distinction has evolutionary significance. The first cells to exist were almost certainly prokaryotic. Current opinion suggests that eukaryotic organisation arose by specialisation of prokaryotes acting as parasites within cells. For example, mitochondria, the small respiratory bodies found in all eukaryotic cells, may have developed from intracellular bacteria.

Micro-organisms are ubiquitous and very numerous. A gram of soil can contain 100 million bacteria and 250 000 fungal cells. Micro-organisms are in the air, in the water we drink, and living within our bodies. A few cause disease in plants or animals, but most are beneficial, particularly as scavengers of organic waste.

5.1 Mycoplasmas are the simplest living cells known; about 60 species are recognised at present. Their DNA codes for about 750 proteins, which is considered the minimum for an independent existence. They differ from bacteria in that they do not possess a cell wall. In this transmission electron micrograph, the mycoplasmas appear as red particles at the surface of an animal cell, coloured yellow. Only a fraction of the animal cell is visible; its diameter is 100 times greater than that of each mycoplasma. The brown line curving across the picture represents the boundary of the animal cell's nucleus. Mycoplasmas can cause pneumonia-like disease in humans and livestock.

TEM, stained section, false colour, ×59 400

VIRUSES

Viruses have been called mobile genes. They exist outside living cells as separate particles called virions. Each virion consists of a length of nucleic acid, the genome, together with a protein coat, the capsid. Some virions are further enveloped in a layer of lipid and protein material. Viruses are classified by shape, size and the nature of their nucleic acid.

The shape of a virus particle is determined by the arrangement of the protein molecules which comprise the coat. These subunits of the coat, called capsomeres, associate together in the form of helices or as the plane faces of polyhedra. Helical viruses may be straight rods, flexuous rods, or bullet-shaped. The commonest polyhedral form is the icosahedron, with 20 faces. The size of a virion depends on the number of capsomeres in its coat, and this in turn is related to the size of the viral genome.

The nucleic acid may be either DNA or RNA. No known virus contains both. The function of the genome is to act as a coding message for the production of enzymes which synthesise the coat protein, and copies of the genome itself. The virus is not alive; to accomplish its own replication it has to enter a living cell. Once inside its host cell, the virion's coat disintegrates to expose the genome, which proceeds to 'hijack' the cell's chemical apparatus to manufacture more coat protein and copies of itself. The new protein and nucleic acid then assemble themselves into the next generation of virions.

The transmission of virus disease can occur through the air, as in influenza, or it may need direct contact, as with rabies. Plant virus diseases are most commonly transmitted by insects with sucking mouthparts.

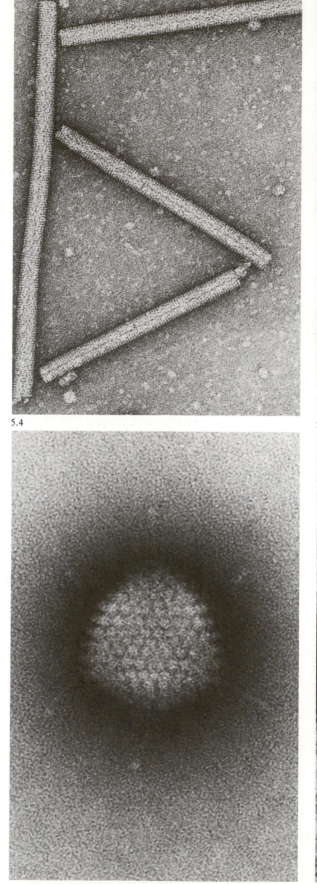

5.2

5.3

5.4

5.5

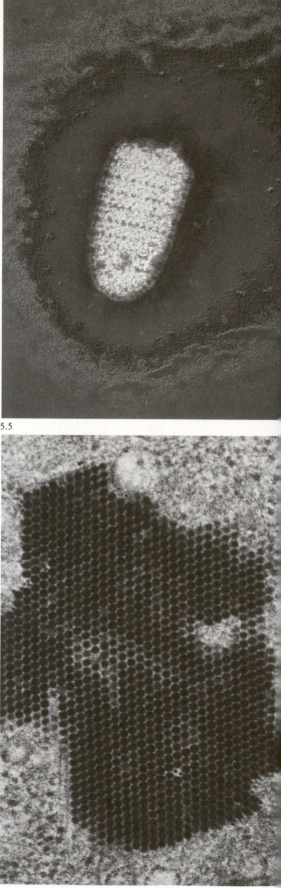

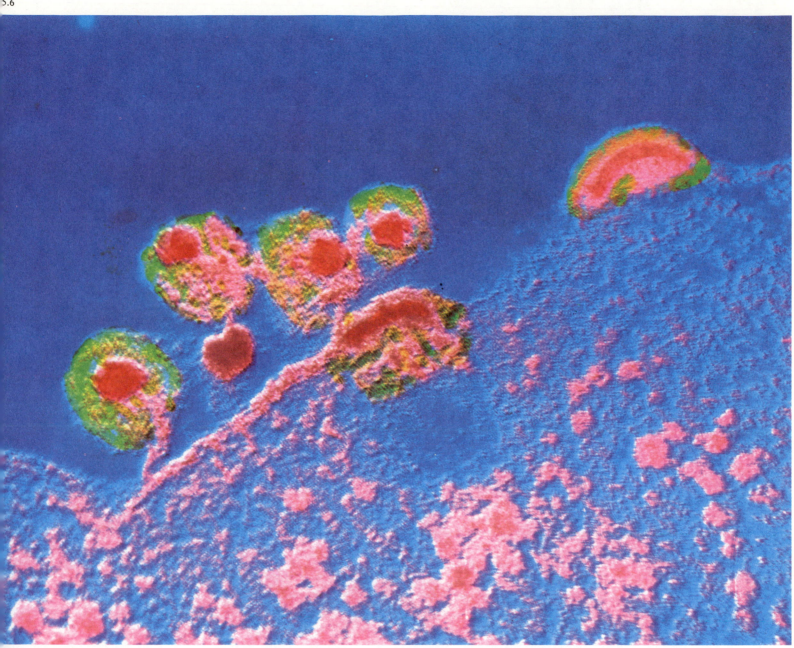

5.2 Beet Necrotic Yellow Vein (BNYV) virus is an example of a rod-shaped RNA plant virus. It is named after the symptoms which it produces on sugar-beet plants – yellowing of the leaf veins, and death. This transmission electron micrograph shows four particles of the virus, each comprising a hollow tube of capsomeres arranged in helical form. The RNA lies inside this tube. The striations show the spacing of the capsomeres, which are 2.6 nanometres apart. This clarity of very fine detail is characteristic of the negative staining technique used to make this high-magnification picture.
TEM, negative stain, ×200 000

5.3 The bullet-shaped rabies virus also has helical symmetry and a genome of RNA. The spiral arrangement of capsomeres can be seen in the light area of the micrograph. The diffuse dark surround is the viral envelope, consisting of lipoprotein material.
TEM, negative stain, ×117 500

5.4 Adenovirus contains DNA enclosed in a protein coat in the form of an icosahedron. The coat is built from 252 protein subunits. The fine spikes visible in this negatively stained preparation enable the virus to recognise the host cell. Adenovirus causes infections of the upper respiratory tract in humans, with

symptoms like the common cold; it has also been implicated in cases of cancer.
TEM, negative stain, ×92 800

5.5 In this section of a cell infected with polio virus, the virus particles are visible in a crystalline array. An infected cell may produce hundreds or thousands of new virus particles. Their dark staining is due to the high concentration of nucleic acid within them.
TEM, stained section, ×83 200

5.6 AIDS (Acquired Immune Deficiency Syndrome) is caused by an RNA virus which uses as its host a white blood cell of the human immune system known as the T4 lymphocyte.

The micrograph shows newly produced AIDS virus particles leaving an infected T4 cell. The deep red core of each particle corresponds to the viral RNA. The new particles will go on to spread the infection, which results in the destruction of the immune response. The victim eventually dies of a secondary infection which his or her immune system is unable to fight.
TEM, stained section, false colour, ×191 000

BACTERIOPHAGES

Bacteriophages are viruses which infect bacteria. They have played an important part in our understanding of genetic processes and in the development of molecular biology. Like other viruses, they consist of a piece of nucleic acid enclosed in a complex protein coat.

The replication cycle of a bacteriophage begins with its attachment to the outer surface of the host bacterium. The nucleic acid within the virus is injected into the host, often by means of a contractile tail. Only the nucleic acid enters the bacterial cell, and this was an important demonstration in the 1950s that DNA on its own is sufficient to encode the structure of complete virus particles. Once inside the cell, the DNA can do one of two things.

In a virulent infection, it immediately starts to direct the synthesis of proteins, including the enzymes concerned with the synthesis of new DNA. Thus the virus circumvents the cell's normal control processes, which ensure that DNA is replicated only once in each generation. With its own enzymes in control, the virus can replicate itself hundreds of times, and within 30 minutes or so the bacterium becomes filled with new virus particles. It then bursts, releasing the new viruses.

In a lysogenic infection, the viral DNA inserts itself into the bacterial chromosome. There it waits, being replicated in the normal way each time the bacterium divides. Thus a single bacterium can give rise to dozens or hundreds of infected progeny, each carrying the virus in latent form. Only when these bacteria are subjected to a life-threatening stress, such as ultraviolet light or ionising radiation, does the viral DNA leave the chromosome and

immediately direct the formation of new virus particles. In this way the virus, but not the bacterium, survives.

Bacteriophages can be used in DNA cloning. A novel piece of DNA is inserted into the viral DNA, and infection allowed to proceed. The host bacterium divides to eventually produce thousands of copies of the inserted DNA. This technique is used for identification and isolation of useful genes, such as those

concerned with the production of antibiotics and human hormones.

Bacteriophages occur naturally in a wide range of environments – in soil, sewage, dairy products and diseased tissue. Unlike the viruses of higher organisms, which are commonly named after the disease they cause, bacteriophages are identified by letters and numbers.

5.7 The complex protein capsule of these two SP105 bacteriophages is in three parts. The large head region

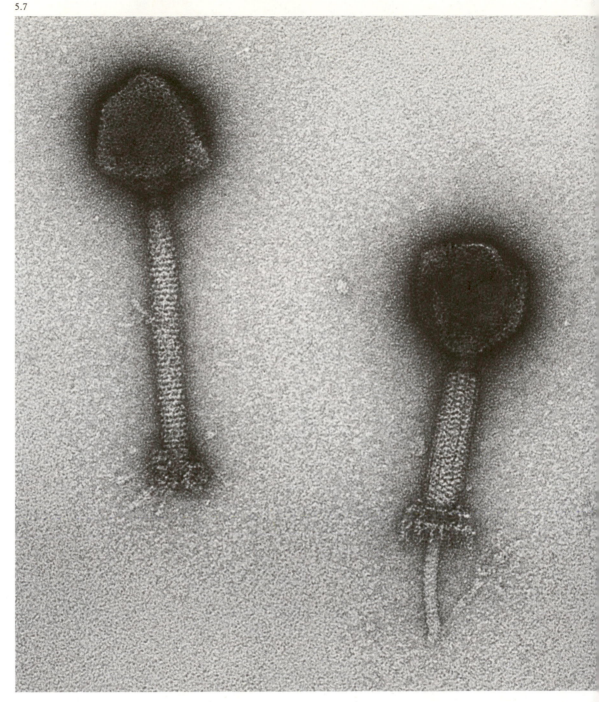

houses most of the DNA. The tail is contractile and serves to inject the DNA into the host through a tubular core. The fine fibres at the end of the tail recognise and bind to the surface of the host bacterium prior to infection. In this transmission electron micrograph, the bacteriophage on the left has its tail in expanded mode, while the one on the right has its tail contracted, revealing the core through which the DNA has travelled.
TEM, negative stain, ×470 000

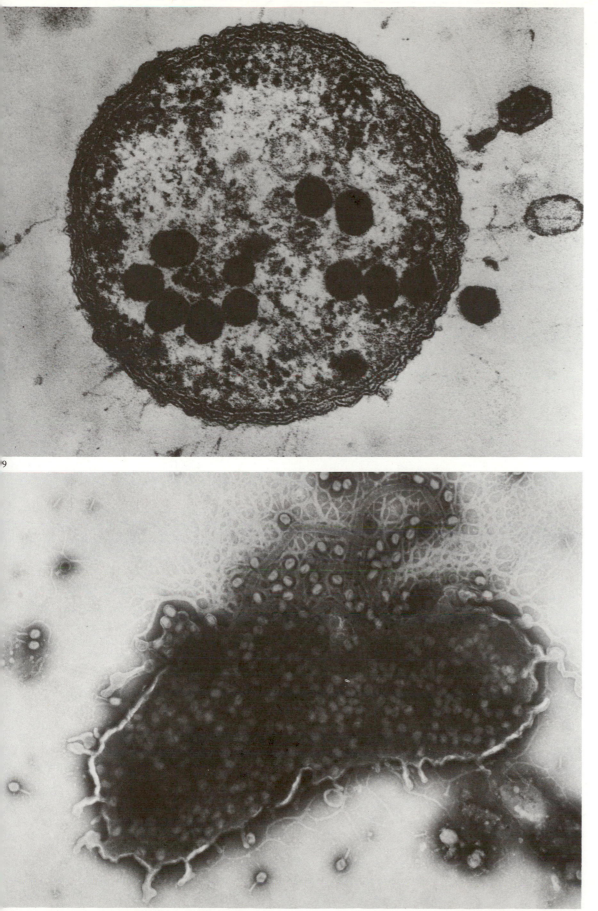

5.8 This section through a cell of the bacterium *Escherichia coli* shows it being attacked by T2 bacteriophages. Nucleic acids stain easily and appear dark in electron micrographs. The empty T2 particle at right has already injected its DNA into the bacterium, and new virus particles are visible as the solid black objects within the cell. The much smaller granular particles in the cell interior are ribosomes, on which protein synthesis takes place. A section like this is very thin – about 70 nanometres thick. This is about 1/50th of the complete bacterial cell.
TEM, stained section, magnification unknown

5.9 The bursting, or lysis, of the host bacterium is the final stage of a virulent infection. Graphically captured in this micrograph, the process results in the release of thousands of new bacteriophage particles. The ones in this picture are T4; they appear as small white particles because the specimen has been negatively stained. The specimen consists of the whole bacterium, not a thin section as in the previous picture.
TEM, negative stain, ×120 000

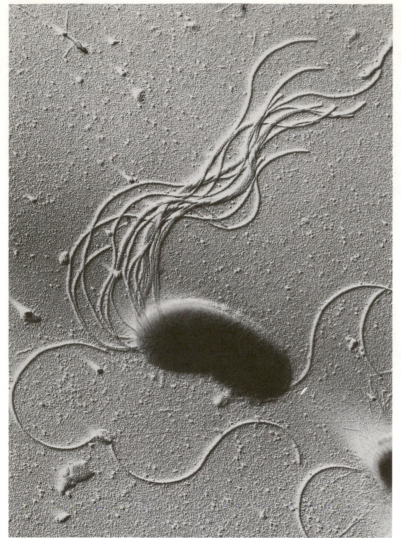

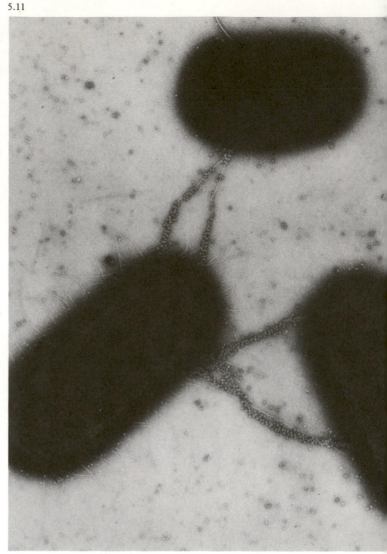

BACTERIA

Bacteria are prokaryotic cellular micro-organisms. There are about 2000 species, ranging in size from spheres 0.3 micrometres across to filamentous cells 20 micrometres long. Bacteria come in three basic shapes, as spheres (cocci), rods (bacilli), or spirals (spirilla). Individual bacteria may associate into groups; the Streptococci, for example, form chains of spherical cells.

The bacterial cell is elementary in structure. It contains DNA in the form of a closed loop as its genetic material. The cytoplasm is filled with ribosomes – the particles on which protein synthesis takes place. The cell membrane may have folded regions, known as mesosomes, on its inside surface. The external surface may be smooth, covered with a slimy capsule, or it may have fine hairs, called pili. These attach the cell to its substrate, and to other cells during sexual union. Motile bacteria have one or more flagellae. These sinuous protein filaments rotate to propel the cell through liquid environments.

Bacteria differ in their ability to use nutrients. The heterotrophs require preformed organic molecules produced by other cell types, and they are therefore associated with other organisms – occurring, for example, in the human mouth and gut. Autotrophs can use inorganic minerals and atmospheric carbon dioxide to satisfy their needs. They are widespread in the soil and water.

A few species of bacteria cause illness, such as the agents of cholera, plague and legionnaire's disease. But the vast majority are beneficial, acting to break down organic wastes, maintain soil fertility, and even digest some components of our food for us.

5.10 To obtain this picture, a cell of the soil bacterium *Pseudomonas fluorescens* was coated with a layer of carbon followed by platinum. This produces a surface replica which is used as the microscope specimen. *P. fluorescens* is motile, and uses the bunch of whip-like flagellae seen here to move through the water layer surrounding soil particles.
TEM, surface replica, ×18 750

5.11 Sex between bacteria, known as conjugation, was discovered in 1946. involves the transfer of DNA between cells. In this picture, three cells of *Escherichia coli* are visible. The one at bottom left is male, the other two are females. Maleness in *E. coli* is associated with the presence of long, hollow surface hairs, called F-pili. The male has attached F-pili to each of the female *E. coli*, and DNA is transferred between them through the hairs' hollow cores. Careful control and interruption of this process enables scientists to map the relative position of the bacterium's genes. The tiny granular particles that almost completely cover the F-pili in this micrograph are particles of MS2 bacteriophage, which bind specifically to the F-pili.
TEM, negative stain, ×10 250

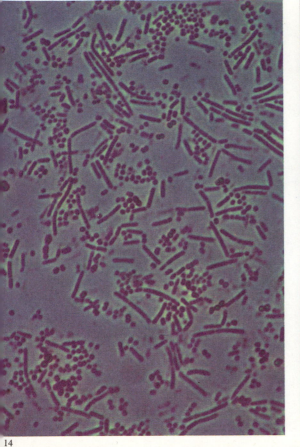

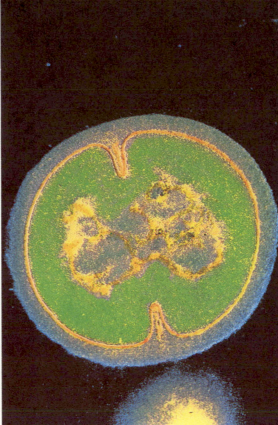

5.12 This light micrograph of a population of bacteria from the human mouth demonstrates the limitation of optical microscopy in the study of such small cells. Compared to the other pictures on these pages, there is very little detail visible beyond the outline shape of the bacteria, a mixture of cocci and bacilli.
LM, bright field illumination, ×1600

5.13 Bacteria divide by splitting in two, a process called binary fission. In this picture of a dividing cell of *Staphylococcus epidermidis*, the blue and yellow region in the cell centre is the genetic material, about to break into two parts. The division is completed by the ingrowth of a new cell wall, seen here as the fine yellow line around the edge of the cell. The blue layer beyond is the external wall covering. *S. epidermidis* is found in quantity all over human skin, where it is generally non-pathogenic, though it may play some part in causing acne and in making cuts and scratches go septic.
TEM, stained section, false colour, ×60 000

5.14 This bacterium, a species of *Leptospira*, takes the form of a long thin spiral. *Leptospira* is one of a group of helical bacteria known as spirochaetes, which contains many dangerous pathogens. The condition leptospirosis is commonly known as Weill's disease, and is transmitted by contact with rats. This single cell, in red, is about 20 times larger than the other bacteria on these two pages.
TEM, negative stain, false colour, ×20 000

14

E. COLI

Escherichia coli, or *E. coli* as it is commonly known, is a normal and usually harmless inhabitant of the intestinal tract. It is also the world's premier laboratory organism. A vast amount is known of its genetics, and many laboratories house large collections of different strains, each capable of carrying out a specific series of reactions. Much of our understanding of how genes work comes from the study of this bacterium.

Today this detailed knowledge is exploited by the techniques of genetic engineering. By deliberately introducing pieces of foreign DNA into the bacterium, it is possible to make it produce useful medical products such as insulin or interferon.

5.15 In the scanning electron microscope, the image is strongly three-dimensional in appearance, and the *E. coli* cells are seen as typical rod-shaped bacilli. The fibrous material is the remnants of the nutrient medium in which the bacteria were grown.
SEM, ×17 000

5.16 By contrast, the transmission electron microscope gives an image of the internal details of cells, with little three-dimensional content. This picture is of a section of *E. coli* only 70 nanometres thick. The black regions are DNA which has duplicated itself in preparation for binary fission.
TEM, stained section, ×78 000

5.17 To make this picture, the bacterial cell in the centre was treated with an enzyme to weaken its wall, then placed in water, causing its DNA to be ejected. It is visible as the gold-coloured fibrous mass lying around the cell, and its length is 1.5 millimetres, or 1000 times the length of the cell from which it came. The specimen has been shadowed with a layer of platinum, a technique which gives very high resolution pictures.
TEM, shadowed replica, false colour, ×67 500

5.15

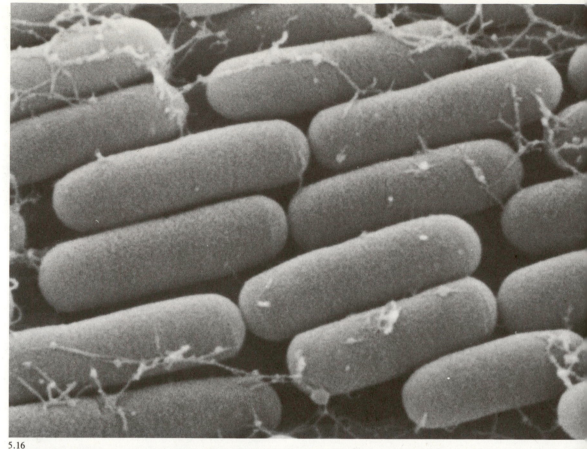

5.16

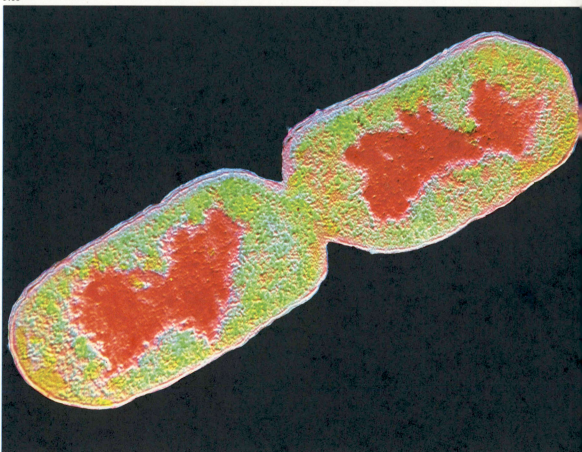

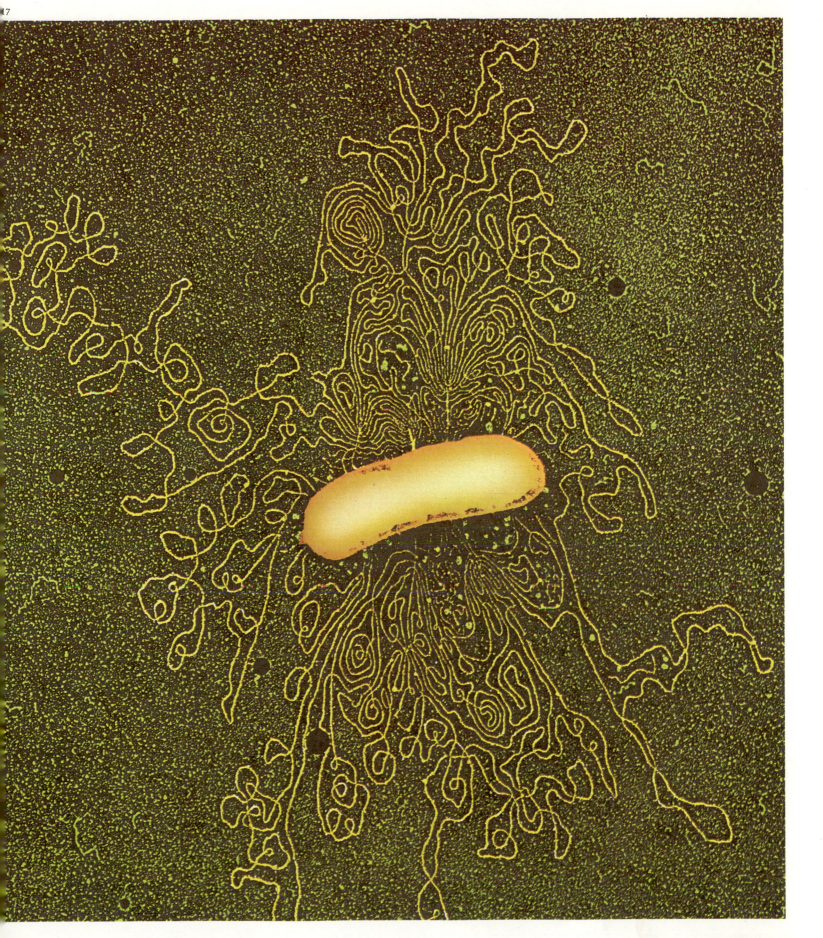

E. COLI 97

RHIZOBIUM

Fertile soils contain many species of bacteria, most of which break down organic materials and create humus. One which has a more directly beneficial effect on plant growth is called *Rhizobium leguminosarum*. It forms a specific relationship with the roots of leguminous plants, a family which includes clover, peas and beans.

The first contact in the relationship is between a bacterial cell living free in the soil and one of the fine hairs which the plant's roots produce in order to extract moisture from the ground. In response to the contact, the root hair admits the bacterium to the interior of the root by means of a specially constructed tunnel called an infection thread. Once inside the root, the bacterium causes the root cells to divide. This eventually produces a large outgrowth from the root surface, called a nodule, which is visible to the naked eye. Within the cells of the nodule, the bacterium loses its outer wall, and divides repeatedly to produce a mass of new cells which, because they lack cell walls, are called bacteroids. Thus the plant acts as an amenable host for the multiplication of a single bacterium into millions of bacteroids.

The relationship, however, is not one of parasitism, but of symbiosis – a partnership beneficial to both sides. In return for its place of residence in the nodule, *Rhizobium* performs a unique service for the plant. In its bacteroid form, it can transform the nitrogen gas in the air which permeates the soil into ammonium salts. These act as nitrogenous fertiliser for the plant.

Rhizobium is by far the most important of the soil bacteria which fix nitrogen for plant use. It has been estimated that 150–200 million tonnes of nitrogen are fixed

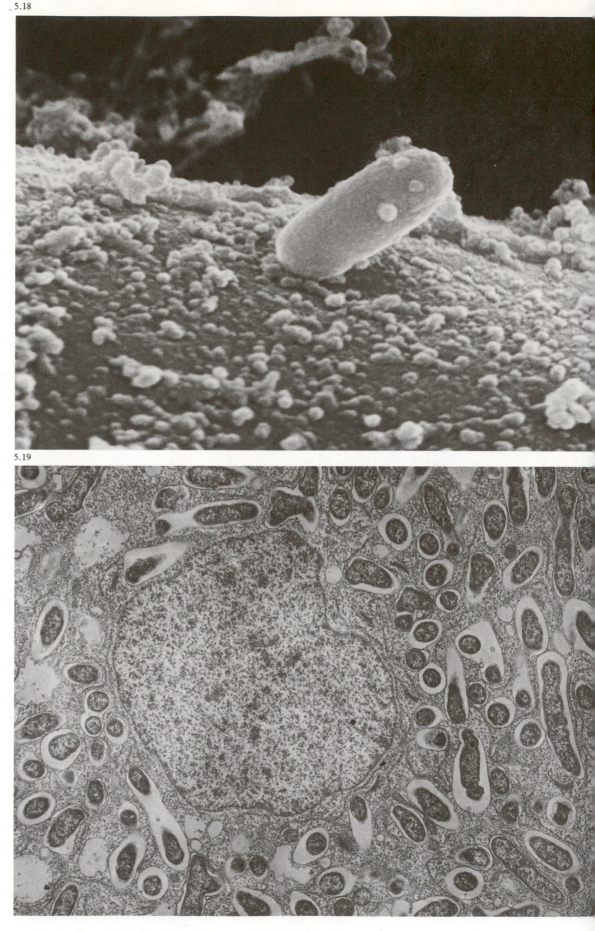

5.18

5.19

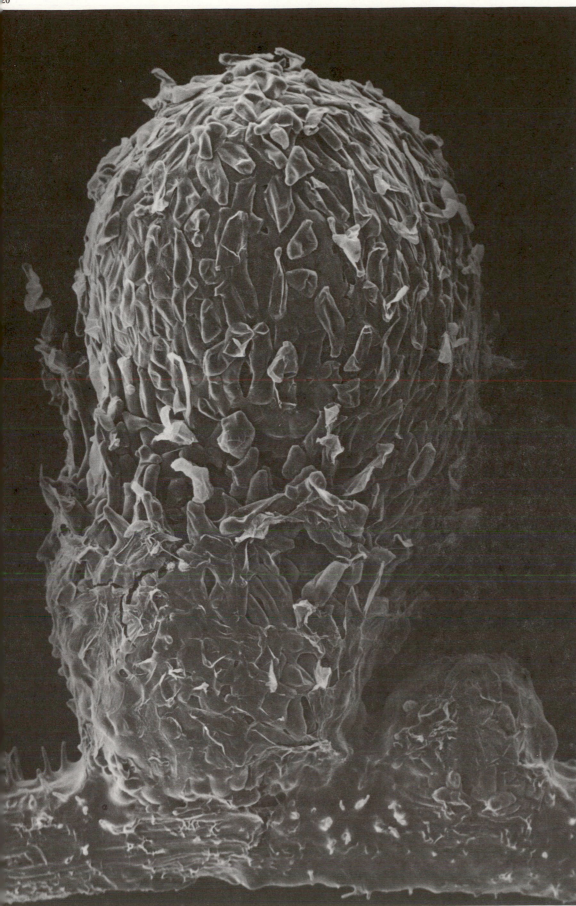

by bacteria annually – about three times the total world production of nitrogen fertilisers from chemical factories. The life cycle of *Rhizobium* is completed when, after a period of a few weeks, the nodule breaks down. The bacteroids then form cell walls and enter the soil as bacteria, to repeat the process of infection.

5.18 This high-magnification scanning electron micrograph shows a single bacterium attached to a root hair of a laboratory-grown garden pea plant. The attachment is very specific: *Rhizobium leguminosarum* infects only leguminous plants, although it is not known how it recognises them; and the initial contact with the root hair often appears to occur at one end of the rod-shaped bacterium, as in this picture. The granular particles on the surface of the root hair and the bacterium are remnants of the medium in which the plant was grown.
SEM, ×40 000

5.19 Each infected cell is host to thousands of bacteroids. In this section, the large central body is the nucleus of a nodule cell. Around it are ranged the small, dark-staining bacteroids; of indefinite shape, they are separated from the host cell by a membrane and a clear space. Half the weight of a mature nodule may consist of bacteroids.
TEM, stained section, ×6250

5.20 This large nodule is about 3 millimetres in diameter. The loosely packed cells on its surface are not normally as infected with bacteroids as internal cells like the one in Figure 5.19. The nodule has grown from the length of pea root at the bottom of the picture. To the left of the nodule's base can be seen a few root hairs; to its right is a much smaller nodule.
SEM, ×35

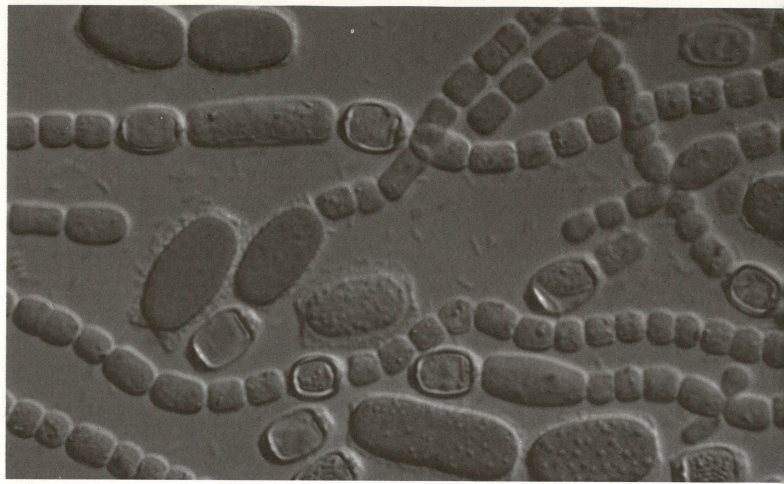

ALGAE

Algae are found wherever there is water and light. They all share the ability to transform carbon dioxide into sugars by the process of photosynthesis. Algae have been called 'the grass of the waters'; they are immensely important as the base of many food chains, particularly in the oceans. Algal marine plankton fixes 10^{10} tonnes of carbon each year by photosynthesis – more than the total production of all the world's land plants.

Most algae are microscopic, existing as single cells or filaments of cells joined end to end. Some, such as seaweeds, form large visible bodies. The life cycles of algae are often complex, with highly specialised sexual reproduction, which takes place in water.

Algae use mixtures of pigments to perform photosynthesis, and this is one basis of a broad classification into classes – the green algae, brown algae, red algae, and so on. The cells of these algae are eukaryotic, with a distinct nucleus and the pigments localised in organelles called chloroplasts. Blue-green algae are primitive relatives, sometimes classified with bacteria. Like bacteria, they are prokaryotic, but their main pigment is chlorophyll, which is found in algae and all higher land plants. It is possible that the chloroplasts of higher plant forms arose originally from blue-green algal cells which were ingested by a non-photosynthetic cell early in evolution

5.21 Different cell types may form even in primitive organisms such as this blue-green alga, *Cylindrospermum*. The chains of small units in the light micrograph are vegetative cells which perform photosynthesis. The slightly larger cells with thick walls are called heterocysts, and they can fix nitrogen. The largest cells are a type of spore called an akinete.
LM, Nomarski DIC, magnification unknown

5.22 The desmids are green algae characterised by their beautifully shaped cell walls. This is a single cell of the desmid *Micrasterias*, seen among a variety of filamentous algae and debris. The cell is in two halves, separated by a narrow waist. Cell division is accomplished by splitting in two at the waist, with each half generating a perfect replica of itself to restore the original shape. The cell in this picture divided some time ago, and the new half cell (upper right) is almost full-grown.
LM, spectral Rheinberg illumination, ×850

5.23 The green alga *Volvox* forms spherical colonies consisting of hundreds or thousands of cells glued together to form a hollow sphere. The individual cells appear as bright green particles in the picture. The larger green masses, of which six are visible, are asexual daughter colonies forming inside the sphere.
LM, dark field illumination, ×315

5.24 *Spirogyra* is a green alga in which the cells join end to end to form filaments. Its chloroplasts are arranged in spiral patterns which are disrupted during cell division. In this picture the cell in the middle of the frame is dividing; those above and below it show the spiral chloroplasts typical of non-dividing cells.
LM, bright field illumination, ×200

5.25 *Hydrodictyon* is another colonial green alga. Colonies can comprise up to 20 000 cells joined so as to make a hollow floating net which is closed at both ends. Large colonies may reach 50 centimetres in length. This picture of a small part of a colony shows how the individual cells join and branch at their ends to produce the three-dimensional network.
LM, Rheinberg illumination, ×735

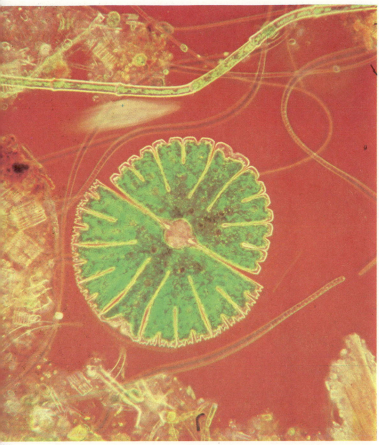

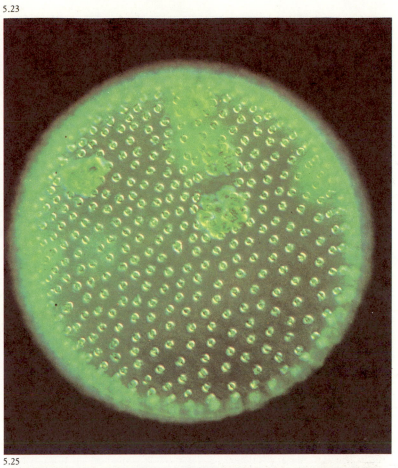

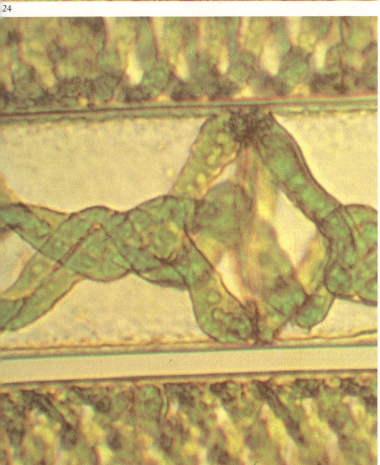

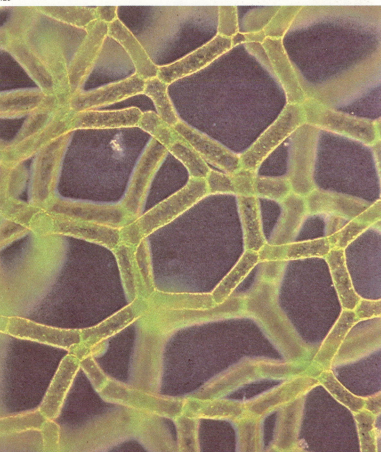

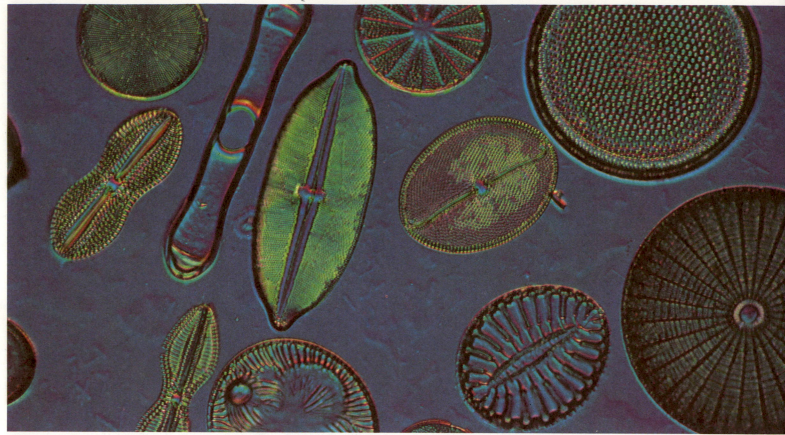

DIATOMS

The diatoms are a distinctive group of single-celled algae. Comprising about 10 000 species, they form an important part of the plankton in fresh and salt waters. The number of diatoms in the seas is immense, especially in temperate latitudes, where a litre of surface sea water may contain as many as 15 000 diatoms.

The characteristic feature of diatoms is their intricately patterned, glass-like cell wall, or frustule. The frustule consists of two halves, the valves, which fit together like the two parts of a pill box. One half is slightly larger than the other, and acts as the 'lid' of the box.

When a diatom divides, each valve produces a new half inside itself. In consequence of this, the 'lid' generates a cell which is identical in size to the original one, whereas the 'bottom' gives rise to a cell which is slightly smaller. This process can be repeated many times, but eventually the ever smaller progeny of the 'bottom' will cease to be viable. The diatom solves this strange paradox by occasionally producing sexual 'auxospores' which restore the maximum cell size.

Diatoms have long been favourite subjects for microscopists. Victorian amateurs amused themselves by arranging diatoms in complex patterns. The frustule, made from silicates which the diatom extracts from the water around it, is often decorated with tiny holes arranged in rows (*striae*) so fine and regular that they have been used as test objects to measure the quality of microscope lenses. In industry, diatomaceous earth formed from skeletons of countless billions of diatoms is used in products as diverse as toothpaste, dynamite and sealing wax.

5.26 Diatoms are classified into two broad groups depending on their structure. In centric diatoms, the striae are arranged radially. In pennate diatoms, the striae are in rows on either side of a central axis of symmetry. The picture shows representatives of both types. The colours are due to the use of polarised light, not to any pigmentation of the frustule.
LM, Nomarski DIC, ×280

5.27 Some diatoms form very simple colonies consisting of cells which have failed to separate following division. This picture shows 14 such cells of the species *Fragillaria crotonensis*. Each one is glued by its centre to its neighbours with a mucilaginous material. Because of the mechanism of division, the cells at either end of this colony are different in size, but the difference is too small to detect with certainty. *F. crotonensis* lives in freshwater lakes.
SEM, magnification unknown

5.28 Many diatoms cannot move independently; those that can usually achieve only a strange jerky motion. *Navicula monilifera* is one such motile species. A pennate diatom, its striae are arranged in rows on either side of a central furrow, the raphe. The raphe the secret of the diatom's mobility. Fluid is squirted along it and pushes the whole cell forwards. The power of movement is most common in bottom living species such as *N. monilifera*; diatoms which float rely on water currents to carry them from place to place.
SEM, ×1250

5.29 The decorated valves of diatoms consist of a mixture of pectin and silicates. In centric diatoms, such as *Cyclotella meneghiniana*, the two valves are separated by an undecorated girdle band. This can expand, allowing the cell to grow. *C. meneghiniana* inhabits the brackish waters of salt lakes and estuaries.
SEM, ×3750

5.30 This cell of *Biddulphia* is lying on its side, and gives a clear view of the girdle band separating the two valves. The band holds the valves together while at the same time permitting cell growth. *Biddulphia* is a component of marine plankton.
SEM, ×1200

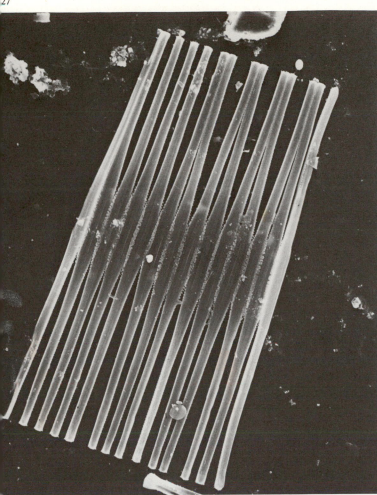

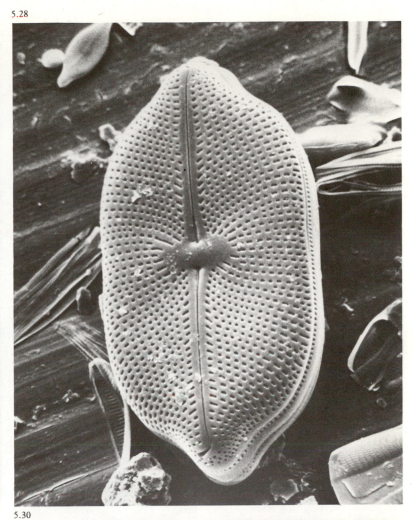

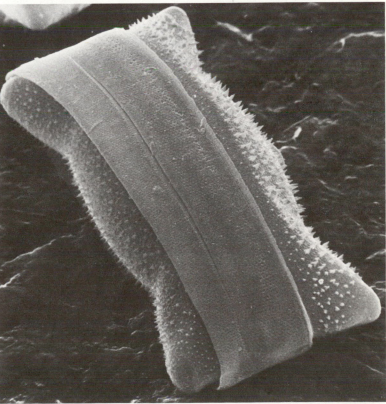

FUNGI

Fungi, like algae, occur in a variety of forms ranging from truly microscopic organisms up to visible and familiar ones such as mushrooms and toadstools. Many exist in the form of single cells. The higher fungi characteristically produce a filamentous growth called a hypha, which branches repeatedly to form a network known as a mycelium. Fungi do not form true tissues. The 'stem' of a mushroom, for example, consists of a tangled mass of hyphae.

The fundamental difference between fungi and algae is the complete absence of chlorophyll in fungi. As a result, fungi cannot perform photosynthesis, and have to rely for sustenance on organic material produced by some other living creature.

Many fungi feed on the dead remains of other life forms. Known as saprophytes, their favourite haunt is the forest floor with its supply of fallen leaves and branches. Mushrooms and toadstools are saprophytes. Saprophytic fungi perform a useful function in the maintenance of soil fertility by contributing to the formation of humus. They are also familiar as the moulds which spoil our food; they are quite at home in a pot of jam, or on the surface of bread, cheese, or an orange. Other representatives live off the fabric of our houses – dry rot and wet rot of timber are both caused by saprophytic fungi.

The other major fungal lifestyle is parasitic: the fungus attacks the living tissue of its host, causing symptoms of disease. It was the impact of a fungus disease – potato blight in 19th century Ireland – which first gave impetus to the science of plant pathology. Fungal diseases of humans are by and large less severe in their effects than bacterial or viral infections.

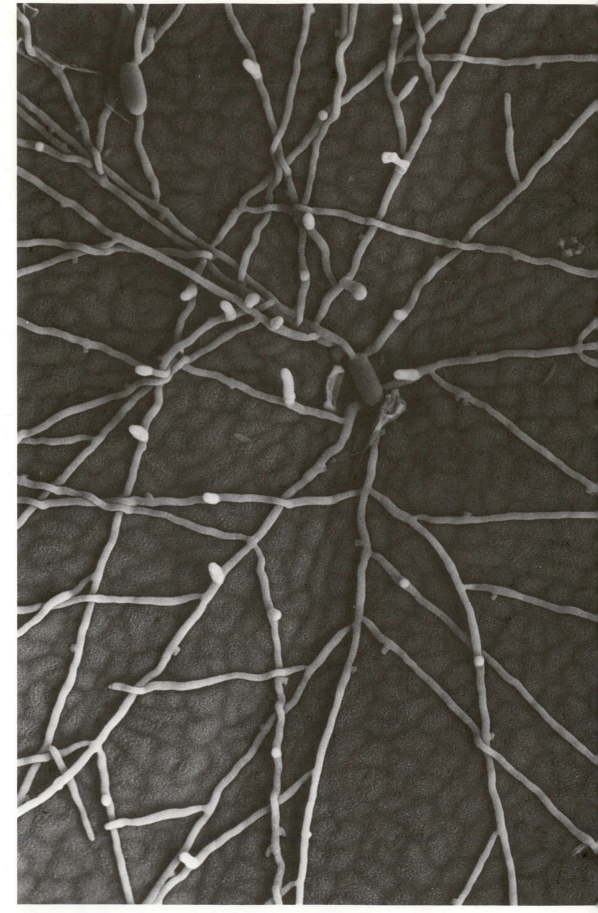

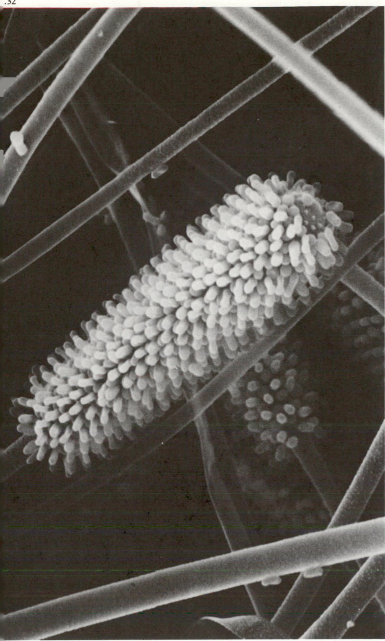

5.34

.31 One of the commonest plant diseases caused by fungi is powdery mildew. This generic name refers to the appearance of an infected plant, the leaves of which seem to have been dusted with powder. The powder in fact consists of the tiny spores of the fungus. A wide range of fungal species are referred to as 'powdery mildews'. The one depicted here is *Erisyphe pisi*, which infects garden peas. The infection begins with a spore landing on a leaf surface. The originating spore can be seen just above centre in this picture. The spore germinates and produces a series of branching filaments, which creep over the leaf surface and at intervals send branches down into the leaf interior. Eventually the fungus develops aerial branches which give rise to more spores. These are blown away by the slightest breeze to spread the infection. Crops such as peas and cereals, and many grasses, are affected by powdery mildews.
SEM, ×280

5.32 The spore-bearing structures (sporangia) of fungi are often elaborate in shape, and can be used as a means of identification. This one, from *Mycotypha africana*, is shaped like a bottle brush. The picture also shows the hyphal filaments of this fungus, which is another powdery mildew.
SEM, ×900

5.33 Skin and hair are rich sources of a protein called keratin. This is used as a food source by a number of fungal species which cause the symptoms of ringworm and athlete's foot. Here, the hyphae of *Trichophyton interdigitalis* are shown growing through the epidermal scales of human skin. The fungus lives in the soil and on small furry animals such as voles; in humans it causes ringworm.
SEM, ×3500

5.34 To see what is happening inside a leaf, it is necessary to break it open. This scanning electron micrograph was obtained by freezing a leaf from a bean plant infected with a rust fungus, *Uromyces fabae*, and then snapping it in half. The view is of the broken edge, and it shows that the interior of the leaf is infested with string-like hyphae of the fungus. Rusts are so called because they produce brown pustules on the surface of infected leaves. These rust spots contain millions of spores ready to spread the infection. A mass of such spores is visible on the surface of the leaf at the top of the picture.
SEM, ×145

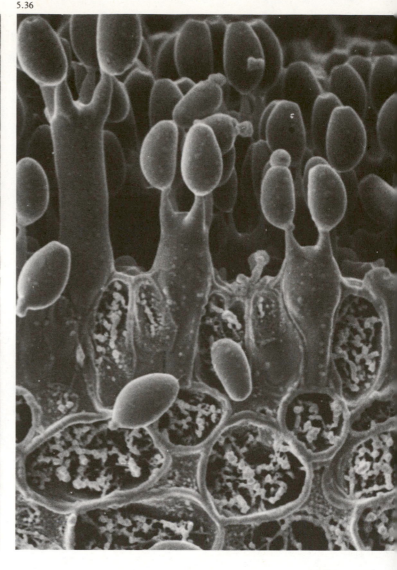

5.35 *Penicillium* species are the most widespread moulds known to humanity: the air is full of their spores, called conidia. As a result, exposed items of suitable foodstuffs such as bread, milk and cheese quickly become infected with the fungus. Initially, growth is in the form of colourless hyphae, but after a few days the fungus produces special aerial branches called conidiophores, seen here growing on a mouldy piece of Cheddar cheese. At the ends of the conidiophores rows of conidia are formed. They are green in colour, and this is why mouldy cheese often looks green. The slightest movement of air causes the conidia to be detached and blown away. It was the chance arrival of a spore of *Penicillium notatum* on a bacterial plate in Alexander Fleming's laboratory that led to the discovery and eventual purification of the antibiotic penicillin. In nature, penicillin is used by the fungus to suppress the growth of bacteria which might compete with it for food, or use it as food. Commercial antibiotic production uses *P. chrysogenum*. Other *Penicillium* species are used in the manufacture of cheeses such as Roquefort and Camembert.
SEM, ×400

5.36 The familiar mushroom or toadstool is only a small part of the life cycle of the fungus which produces it. It results from years of growth of the mycelium in the soil. When conditions are right, usually in the autumn in temperate regions, parts of the mycelium associate together to form a compacted tissue mass which lifts itself out of the ground. This 'fruit body' – the visible mushroom or toadstool – produces millions of special cells called basidia from a surface layer known as the hymenium. In this scanning electron micrograph of the fruit body of *Coprinus disseminatus*, the basidia are the elliptical cells at the top of the picture. The bottom of the picture shows a cross-section of the hymenial layer. Sexual union takes place inside the basidia, and this results in the formation of small cells called basidiospores, which are released and carried away by air currents to found new colonies of the fungus. After release of the spores, the fruit body breaks down and disappears, but the mycelium remains alive and will produce more fruit bodies in subsequent years. *C. disseminatus* is known as the 'crumble cap'. It is edible, but very small, and grows on rotting tree stumps.
SEM, ×4000

5.37 Brewer's yeast, *Saccharomyces cerevisiae*, is a single-celled fungus which divides by budding. Several small buds are visible on the yeast cells in this scanning electron micrograph. Yeast has been used in the production of alcoholic drinks for nearly 5000 years. The alcohol is produced as a result of the yeast feeding on sugar in the absence of air. In beer-making, the sugar comes from the germinating grain of barley. Wines are produced using a different species of yeast, *S. ellipsoideus*, which grows naturally on the skin of grapes. Yeast is also used in bread-making. Within kneaded dough there is plenty of air, so the yeast does not produce alcohol. Its respiration results, however, in the formation of carbon dioxide gas, and it is this which causes the dough to rise. Yeasts are grown commercially and harvested for the production of vitamin B1, riboflavin and nicotinic acid. Under favourable conditions of culture, a yeast cell can turn into two in the space of about 100 minutes.
SEM, ×2550

o force of any

ons shall inform all

Nations and accepted
stitutional processes. 3.
other States Parties still

ted Nations for

CHAPTER 6
THE CELL

A LL living things are made from fundamental units called cells. A cell is a compartment bounded by a membrane – the plasma membrane – which controls the flow of materials between the cell and its environment. Many organisms consist of single cells – for example, the protozoa. Larger creatures contain thousands, millions, or billions of cells. Just one drop of blood from a pricked finger contains about five million cells.

The word cell was coined in its scientific use by Robert Hooke, to describe the compartments in a slice of cork. Although light microscopy is adequate for studying whole cells, it was not until the invention of the electron microscope – and techniques for cutting very thin sections – that the complexity of their internal structure was appreciated. Cells are classified into two types, prokaryotic and eukaryotic, depending upon their internal organisation. This chapter is concerned with eukaryotic cells: those whose genetic material is concentrated in a distinct *nucleus*.

The interior of the cell, bounded by the plasma membrane, is filled with a fluid called cytoplasm. Within the cytoplasm are various bodies, known as cell organelles, which are also enclosed by membranes. Principal of these is the nucleus, with its content of genetic material. All eukaryotic cells also contain mitochondria, which are organelles concerned with energy production and respiration. Plant (but not animal) cells contain a family of organelles known as plastids, of which one member, the chloroplast, is essential for photosynthesis. All cells also contain a variety of other membranes in the cytoplasm,

collectively called the endomembrane system. These are concerned with the chemical synthesis of polymers such as proteins, and their transport within the cell.

The fluid of the cytoplasm was formerly regarded as a more or less uniform mixture of soluble chemicals peppered with tiny particles, the ribosomes. It is now known to be permeated with a network of protein filaments, which are thought to act as a 'skeleton' for the ordered movement of chemicals and organelles.

During growth, the number of cells in an organism increases by cell division. After division, cells usually adopt a specialised function. In the human body, about 200 different cell types can be distinguished on the basis of their structure and biochemical speciality. In plants the number of different cell types is about 20.

6.1

6.1 One of the characteristics that distinguishes plant cells from animal cells is that plant cells have an additional external envelope outside the plasma membrane. Made from a mixture of proteins and polysaccharides, including cellulose, it is known as the plant cell wall. In this transmission electron micrograph of a group of cells inside the root tip of a maize plant, *Zea mays*, the wall appears as a thin layer between the cells. The wall defines the shape of the cell, and it expands as the cell grows. The prominent round organelle in each cell is the nucleus; and each nucleus, in turn, contains a smaller, dark-staining body called the nucleolus. The white areas in the cytoplasm are vacuoles – water-filled spaces which expand and coalesce during cell growth. The grey bodies in the cytoplasm are mitochondria (pale grey) and plastids (darker grey). The small white ovals within the plastids are starch. These particular root cells form part of the meristem – the region which functions to produce new cells continually as the root elongates. Their 'speciality' is to divide; similar cells are found in shoot tips and under the bark of woody stems, such as the trunk and branches of trees.
TEM, stained section ×5000

6.2 Unlike plant cells, animal cells do not have a rigid wall; their plasma membrane does not impose a fixed shape on the cell. In this section through a human lymphocyte, the cytoplasm is coloured green. The large nucleus (orange-brown) occupies most of the internal space of the cell. The orange spot near the bottom of the nucleus is the nucleolus, while the darker brown material around the inside edge of the nuclear membrane is the genetic material, called chromatin. The smaller organelles in the cytoplasm, also orange-brown in colour, are mitochondria. Lymphocytes are part of the immune system, and their function is to produce antibodies. This involves the activation of the cell, which expands from its resting state shown here.
TEM, stained section, false colour, ×26 000

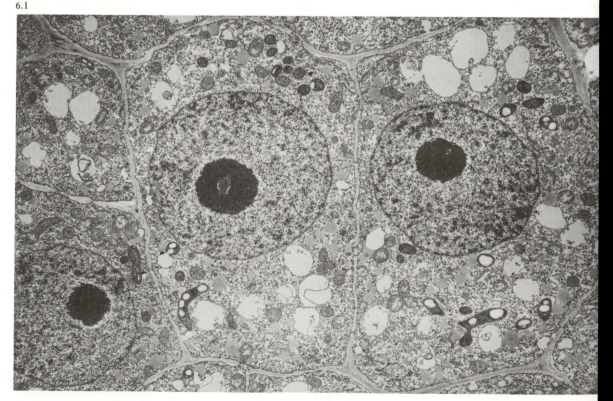

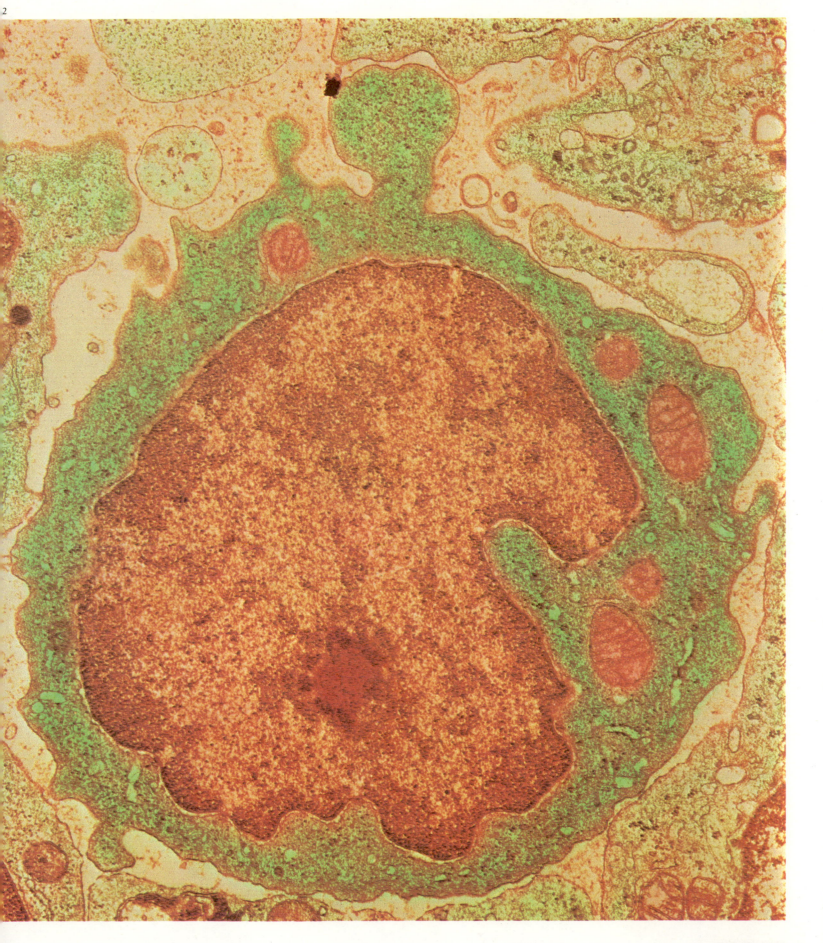

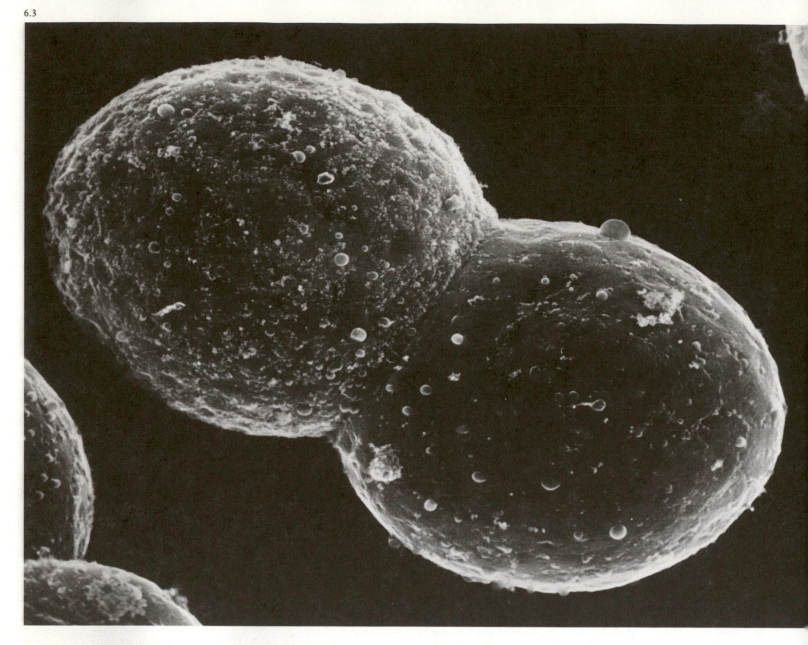

6.3 A plant cell within its wall is like a balloon inside a box: its shape is determined by the box, and it is protected from external damage. If the cell wall is dissolved away, the cell becomes spherical. In this artificial state it is called a 'protoplast'. Protoplasts are very delicate and have to be kept in a special nutrient medium to prevent their bursting, but they have several properties of great interest to biologists. One is their ability to reform their cell walls and, in certain cases, to grow back into normal plants. Another is the fact that, because their plasma membrane is exposed, protoplasts can be fused together by means of chemicals or electrical impulses. This scanning electron micrograph shows two protoplasts, derived from leaf cells of a tobacco plant, *Nicotiana tabacum*, which are in the act of fusing together as a result of treatment with the chemical, polyethylene glycol. The final fusion product will be a single cell which contains two sets of genes, one from each protoplast. This is the reason for the interest in protoplast fusion: if two protoplasts from different plants can be fused together, and a new plant grown from this fusion product, the result will be a hybrid. And because protoplast fusion can be artificially triggered, it is not subject to the natural barriers which limit other methods of plant breeding. It is theoretically possible, therefore, to create entirely new hybrid plants. Although protoplast techniques are comparatively new – the first experiments were performed in the late 1960s – some advances have been made in this direction. Another advantage of protoplasts is that they present plant breeders and genetic engineers with a means of handling large numbers of plant cells in uniform solution. This allows experiments to be carried out – for example, on resistance to virus disease – which would be impossible using whole plant tissues.

SEM, critical point dried specimen, ×2100

The nucleus is the site of most of the cell's genetic material or DNA. Nuclear DNA is always associated with proteins to form a complex called chromatin which exists in paired pieces, each of which is known as a chromosome. The number of chromosomes is determined by species – human cells, for example, have 46 chromosomes in 23 pairs. When a cell is not dividing, the DNA is in an extended form and not usually visible as discrete chromosomes.

The DNA's function is to tell the cell what to do. The chemistry of the cell is controlled by proteins called enzymes, which act as catalysts for chemical reactions. And the structure of each enzyme is encoded by an individual piece of DNA, a gene. A chromosome may contain many thousands of genes, but not all of them are directly concerned with the coding of enzyme structure.

The action of the genes in directing the synthesis of enzymes is extremely complex and involves production from the DNA of different forms of the related nucleic acid, RNA. The RNA passes out of the nucleus and takes part in the job of enzyme synthesis in small particles in the cytoplasm called ribosomes. The ribosomes are themselves made of protein, manufactured in the cytoplasm, and RNA which comes from the nucleus. Thus the activity of the nucleus necessitates a continual export of material into the cytoplasm. This export occurs through pores in the envelope which surrounds the nucleus.

6.4 This view of the external surface of a nucleus was produced by freezing the cell, fracturing it in a vacuum, and then coating the specimen with metal. The micrograph is of the metal 'replica'. It shows part of a nucleus

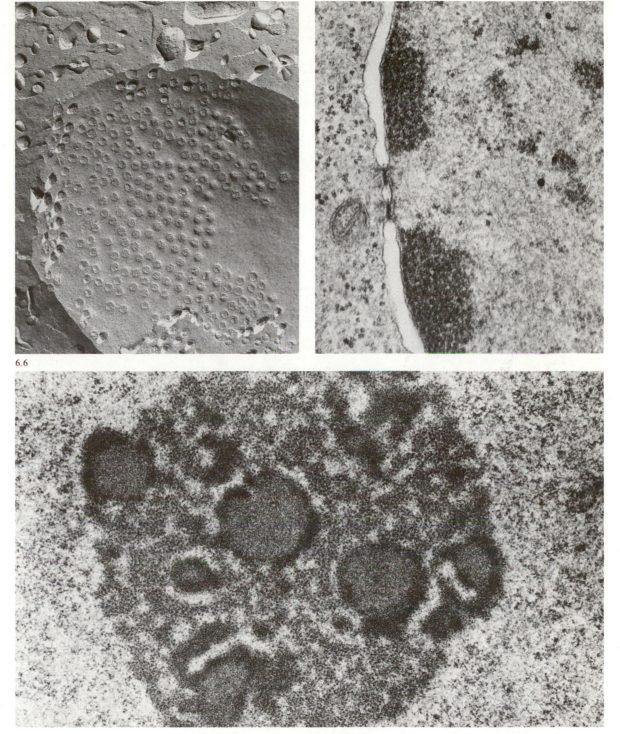

6.4

6.5

6.6

together with surrounding cytoplasm. The nuclear envelope is randomly studded with circular pores; some areas lack pores entirely. At the bottom of the picture, the envelope has split, revealing the inner membrane layer.
TEM, freeze-fracture replica, ×16 000

6.5 In this high-magnification detail of a sectioned cell, the nuclear envelope is seen edge-on, separating the nucleus at right from the cytoplasm at left. The envelope consists of a pair of membranes, separated by the white space. Two pores are visible, each closed by a dark-staining diaphragm. The large, dark, granular masses hugging the inside of the nuclear envelope are chromatin.
TEM, stained section, ×50 000

6.6 The nucleolus is the region of the nucleus where genes which specify the structure of ribosomal RNA are active. In this section through a nucleolus, the dark-staining material is RNA which will form into ribosomal particles within the cytoplasm. The very fine granular regions are sites at which RNA synthesis is occurring on the ribosomal RNA genes. The coarser granular material is ribosomal RNA awaiting export to the cytoplasm.
TEM, stained section, ×25 000

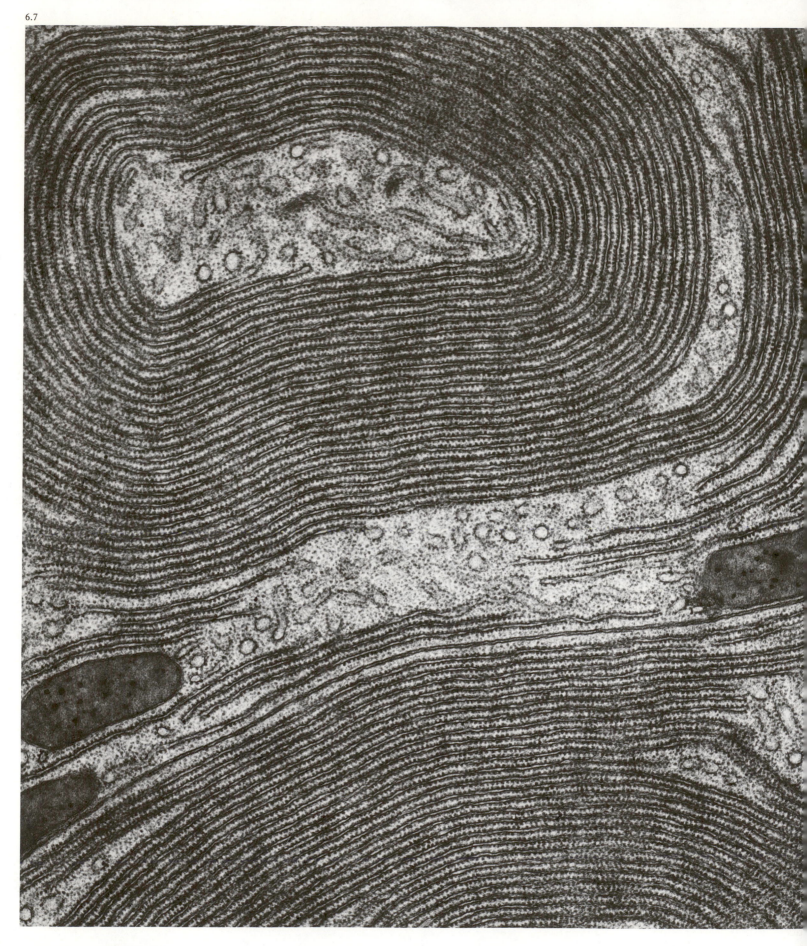

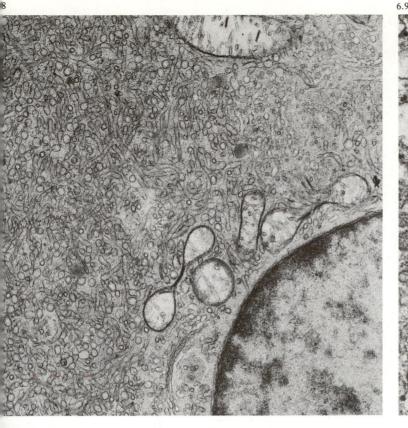

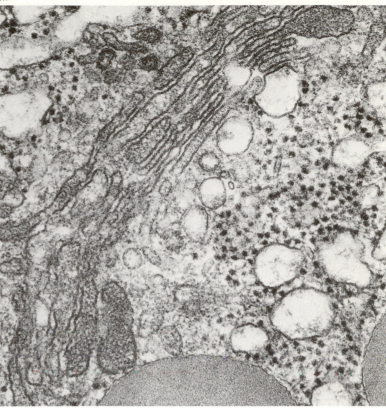

ENDOMEMBRANE SYSTEM

The membrane system of the cytoplasm has two main components – the endoplasmic reticulum and the Golgi apparatus. The endoplasmic reticulum occurs in two forms. *Rough endoplasmic reticulum* consists of extended sheets of paired membranes, and the external surfaces of these membranes are covered with ribosomes. This type of endoplasmic reticulum is found in cells actively engaged in protein synthesis. The chemical process of protein synthesis takes place on the ribosomes, but further processing of the proteins, and their movement to other sites in the cell, can take place within the enclosed space – the lumen – between the paired membranes. Examples of cells which have well-developed rough endoplasmic reticulum are those producing protein hormones in animals, such as the insulin-synthesising cells of the pancreas, or storage proteins, such as cells in developing leguminous seeds of peas and beans.

The other type of endoplasmic reticulum is called *smooth endoplasmic reticulum*, because it lacks ribosomes. It exists in the form of tubes, or as distended, paired sheets of membrane, and is found in cells which are engaged in the synthesis of lipids, steroids, or other hydrophobic polymers. The synthesis of these materials takes place within the lumen of the endoplasmic reticulum; again, products may move about the cell within this enclosed space.

The Golgi apparatus, named after an Italian count, consists of a series of membranous organelles called dictyosomes. Each dictyosome is a stack of paired membranes, and it acts as a sorting office, packaging the products of the endoplasmic reticulum and directing them to their correct destination. Materials enter the dictyosome at one end of the stack of membranes – called the forming face. They leave the dictyosome wrapped up in small spherical vesicles which are budded off from the other end of the membrane stack – the maturing face. On their passage through the membrane stack, they may be chemically modified in order to ensure their correct arrival at their ultimate destination.

The Golgi apparatus is particularly well developed in cells which are exporting a chemical product. Thus dictyosomes are common and active in cells producing digestive enzymes in animals, and they are also found in large numbers in plant cells which are growing rapidly and therefore producing large amounts of wall materials. The vesicles fuse with the cell's plasma membrane and discharge their contents into the extracellular space.

6.7 This micrograph is of rough endoplasmic reticulum in a cell from the pancreas of a bat. The paired membranes appear in large whorls, with tiny black ribosomes attached to their outside surfaces. The space within the endoplasmic reticulum membranes is continuous; it is also separate from the cytoplasm. This cell is engaged in the synthesis of digestive enzymes; the total surface area of endoplasmic reticulum in such cells exceeds 1 square metre for every cubic centimetre of cell volume.
TEM, stained section, ×31 000

6.8 This cell, from the testis of an opossum, is engaged in steroid synthesis. The cytoplasm is packed with smooth endoplasmic reticulum in the form of branching tubes. The large round body at bottom right is the cell nucleus; the other organelles visible are mitochrondria.
TEM, stained section, ×15 000

6.9 This section through a single dictyosome shows the forming face at lower left and the maturing face at top right. The dictyosome consists of a stack of six paired membranes curving diagonally across the picture. Within the lumen of each paired membrane appears a grey, granular deposit – the protein destined for export. The white areas, bounded by a membrane and surrounded by the black dots of ribosomes, are fragments of rough endoplasmic reticulum. The large grey bodies at bottom are droplets of lipid.
TEM, stained section, ×30 000

MITOCHONDRIA

Mitochondria are found in all eukaryotic cells, plant and animal. They are the site of cell respiration – the chemical process which uses molecular oxygen to oxidise sugars and fats in order to produce energy. The energy is stored in the form of a small molecule, adenosine triphosphate or ATP, which is used throughout the rest of the cell to drive chemical reactions such as those involved in forming new proteins, and to power movement within the cell.

Mitochondria are about the same size as bacteria, and they are just visible in the light microscope. Optical examination shows them as small thread-like structures in living cells, continually changing shape and moving in the cytoplasm.

The mitochondrion is bounded by a double-layered membrane. The inner membrane is folded extensively to produce ingrowths called *cristae*. The cristae are where the complex reactions of respiration take place. The mitochondrion's internal fluid, the 'matrix', contains large numbers of ribosomes, together with a small amount of DNA. Using this DNA, the mitochondrion is able to make some of the proteins it requires to carry out respiration. The remainder are imported from the surrounding cytoplasm.

A cell may contain dozens or hundreds of mitochondria. When it divides, its mitochondria are partitioned between the two daughter cells, each receiving roughly half the total. To maintain their numbers through repeated cell divisions, the mitochondria themselves divide. This is accomplished by a process of binary fission similar to the division of bacteria.

In certain environments, mitochondria develop a fixed

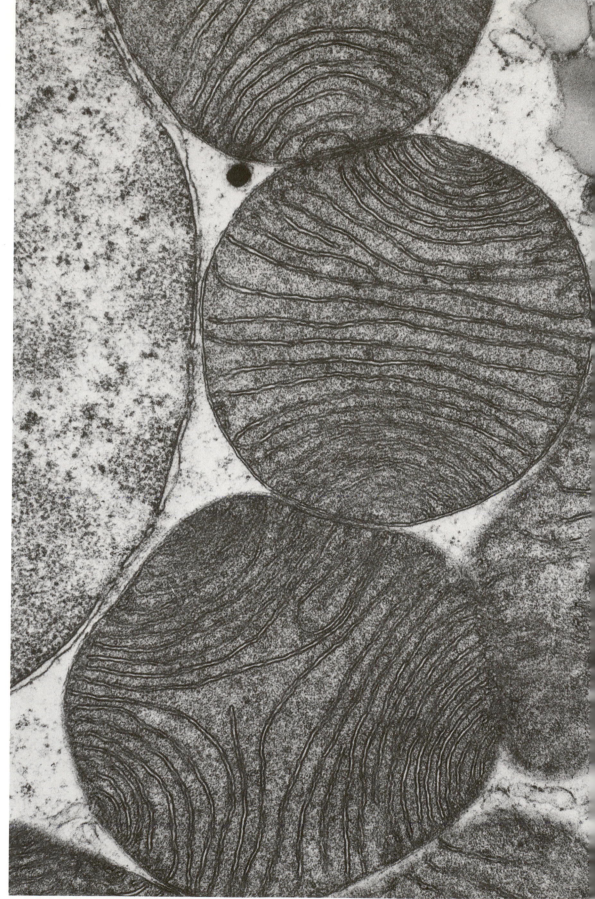

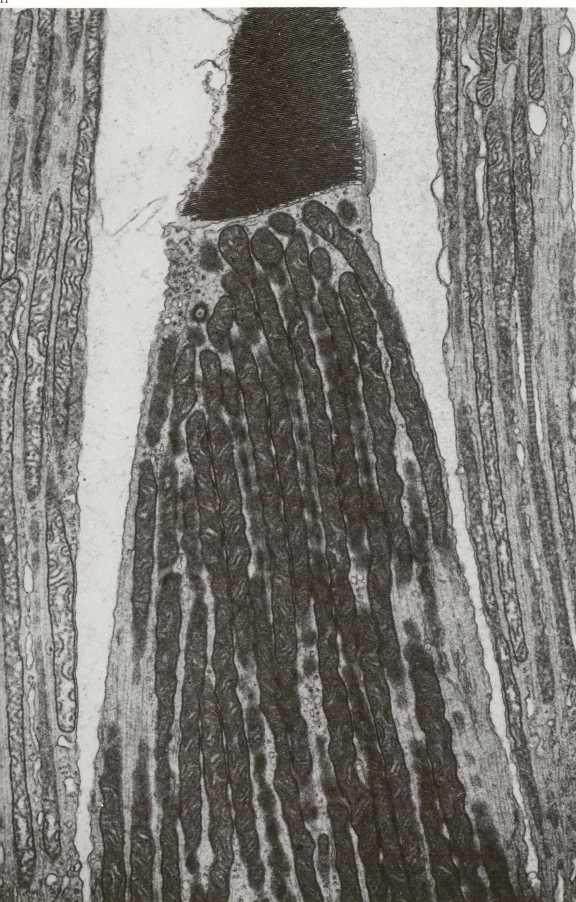

relationship to other cell components. In heart muscle, for example, they are always found close to the muscle fibres, and in sperm they are wrapped round the tail. These arrangements allow ATP produced by respiration to reach its site of action as quickly as possible.

Mitochondria are thought to have arisen by the modification of intracellular bacteria early on in the evolution of eukaryotic cells. They are similar to bacteria in size, they divide like bacteria, and their ribosomes are poisoned by some antibacterial antibiotics. If this idea is correct, it demonstrates an extremely subtle form of symbiosis; mitochondria depend on the rest of the cell for many of their proteins, and the cell depends on the mitochondria for its energy.

6.10 These mitochondria in a brown fat cell from a hibernating bat are particularly large, about 5 micrometres in diameter. Their cristae form narrow bands extending right across the matrix of the organelle. To the left of the picture, part of the cell nucleus is visible. These mitochondria generate the heat necessary to arouse the bat from its hibernation.
TEM, stained section, ×43 000

6.11 The human retina has two types of photoreceptive cells, called rods and cones. This transmission electron micrograph shows a section through a cone cell (centre) and part of two rods (left and right). Both cell types have very elongated mitochondria; those of the cone cell are stained more densely. The black, very fine and closely stacked layers, or lamellae, at the top of the cone cell are the site of the photosensitive pigment. The process of photoreception is poorly understood, but the development of such striking arrays of mitochondria suggests that it requires large amounts of energy.
TEM, stained section, ×3900

CHLOROPLASTS

Chloroplasts are membrane-bound organelles which occur in algae – except the blue-green algae – and in green tissues of all higher plants. They are the sites of photosynthesis, and contain the green pigment chlorophyll in a highly ordered arrangement within their internal membranes. Light falling on the chlorophyll releases electrons which initiate a complex chain of chemical reactions leading to the formation of sugar (sucrose) from carbon dioxide. The sugar may be stored within the chloroplast as starch, or it may be exported throughout the plant via the phloem of the vascular network. Molecular oxygen – the 'breath of life' to ourselves and other animals – is a waste product of photosynthesis.

Chloroplasts are larger than most mitochondria and typically lens-shaped, with a long axis of about 5 micrometres. They are bounded by two membranes; the inner one gives rise to a series of interconnected stacks called *grana*. The interconnections between the grana are known as 'frets'. The fluid within the chloroplast, the *stroma*, contains ribosomes, regions of DNA called nucleoids, and sometimes fat droplets or starch grains. Like mitochondria, chloroplasts are able to make some of their own proteins, but not all; and they divide by binary fission.

Chloroplasts are one member of a family of organelles called plastids, which have different functions depending on the type of cell in which they occur. Photosynthetic cells have chloroplasts, but in storage tissues such as the potato tuber, the plastids exist as amyloplasts, with large starch grains. In carrot roots and the petals of yellow flowers, the plastids are chromoplasts, containing yellow pigment molecules.

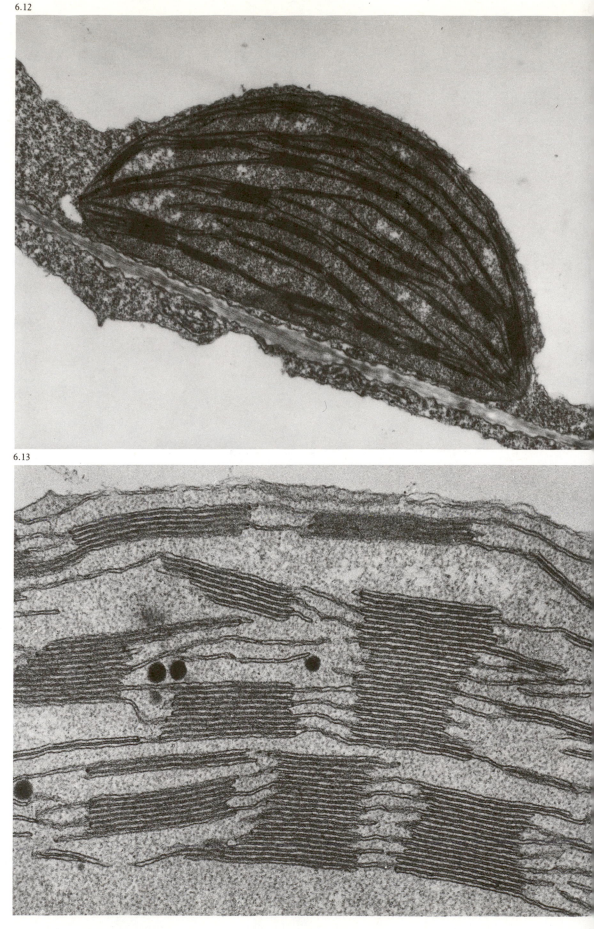

6.12

6.13

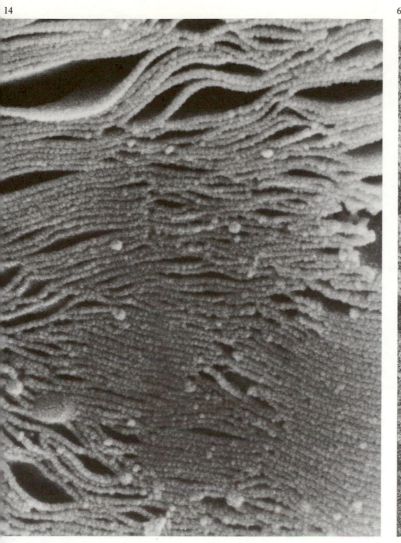

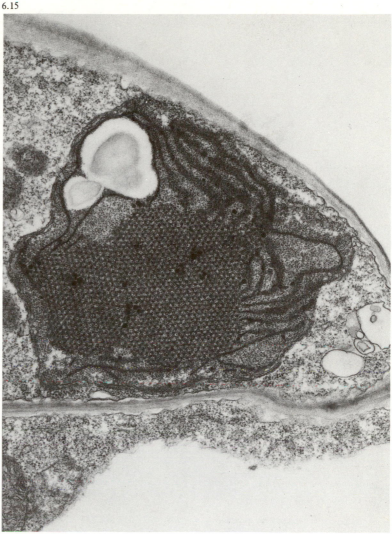

The number of chloroplasts in a cell is variable. Some algae possess only one per cell, whereas the leaves of higher plants usually contain 20–50 per cell. This means that every square millimetre of leaf surface exposes about 500 000 chloroplasts to the light.

As with the origin of mitochondria from bacteria, chloroplasts are thought to have started as intracellular prokaryotic cells, similar to the present-day blue-green algae. Genetic evidence shows that the protein-synthesising machinery of the chloroplast resembles that of prokaryotes.

12 In this transmission electron micrograph of a chloroplast in a leaf cell of tobacco, *Nicotiana tabacum*, the lens shape is clearly demonstrated. The grana appear as closely packed membranes connected to each other by a number of single-membrane frets. The paler regions in the stroma are nucleoids, where the chloroplast DNA is situated.
TEM, stained section, ×27 700

6.13 At higher magnification, the stacked structure of the grana is more obvious. The reason for this formation is not clear, since photosynthesis can occur in plants which do not form granal stacks. The black particles in this picture of a leaf chloroplast of maize, *Zea mays*, are fat droplets. They act as a reserve of raw materials for the production of new membranes.
TEM, stained section, ×43 500

6.14 An entirely different view of the inside of the chloroplast is given by the scanning electron microscope. This picture shows the membranes comprising the granal stacks edge-on. The granular appearance is due to the method of preparing the specimen, which is a dried and broken chloroplast from the spotted laurel, *Aucuba japonica*. This is a very high-magnification picture for scanning electron microscopy.
SEM, fractured dried specimen, ×62 500

6.15 Chloroplasts of flowering plants need light for their formation. A seed germinated in darkness produces pale spindly growth, and such a plant is said to be etiolated. The leaves of etiolated plants contain special plastids called etioplasts. The one in this picture is from a seedling of maize, *Zea mays*. Instead of granal stacks, its membranes have formed a crystalline array, with a few frets. Normal chloroplast structure is generated from this crystalline array within a few hours of an etiolated plant being exposed to light. The white body within the etioplast is a starch grain.
TEM, stained section, ×31 500

CYTOSKELETON

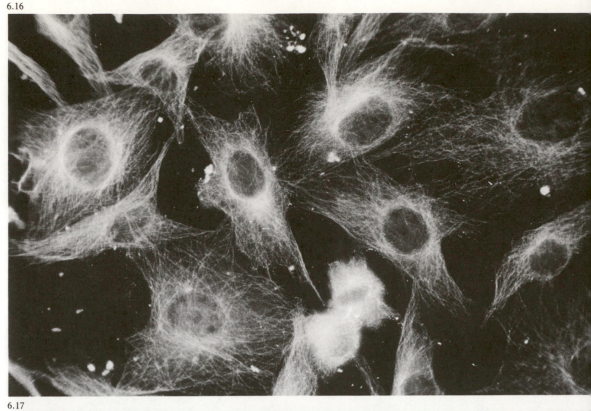

The cytoplasm of cells is highly organised. The smaller organelles are in continuous motion, and the chemical products of metabolism are distributed precisely to different sites within the cell. The basis of this organisation is poorly understood, but advances in light microscopy have shown that the cytoplasm is permeated by networks of tubular and filamentous proteins. These three-dimensional arrays of protein molecules are collectively known as the cytoskeleton.

The components of the cytoskeleton fall into two groups. The *microtubules* are rigid protein tubes assembled from subunits. They are thought to act as direction markers within the cell – a sort of cytoplasmic railway. The filamentous components of the cytoskeleton (*microfilaments* and *intermediate filaments*) form three-dimensional networks, and are thought to be directly involved in the movement of organelles, vesicles and membranes, probably by a sliding mechanism similar to that found in muscle fibres.

The arrays can be visualised by staining the protein molecules with antibodies coupled to fluorescent dyes, and viewing the cell in an ultraviolet fluorescence microscope. It is likely that all cells contain a cytoskeleton, but because of technical difficulties associated with the plant cell wall, most work has been carried out on animal cells.

6.16 This picture, of a group of animal cells which have been grown in liquid culture, shows the microtubule component of the cytoskeleton. The cells have been fixed and stained with an antibody which binds to microtubules. The antibody has been chemically attached to the dye fluorescein, which produces an intense green fluorescence in ultraviolet light.

6.17

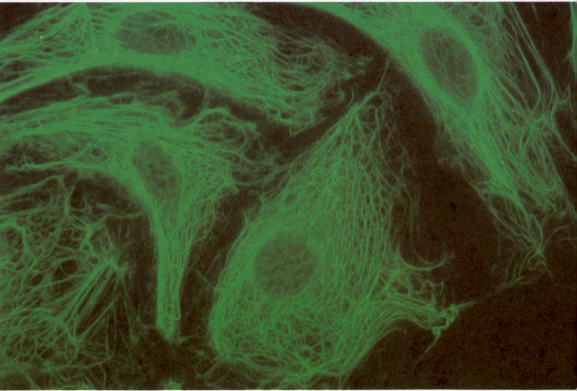

The large, apparently empty space in the centre of each cell is the nucleus. The microtubules radiate out from the nucleus to the cell periphery. They are predominantly straight.
LM, ultraviolet fluorescence, anti-tubulin stain, ×200

6.17 This micrograph shows the distribution of an intermediate filament protein called vimentin. The cells are from the kidney epithelium of the kangaroo-rat; they have been grown on a glass cover slip, fixed, and stained with antibody to vimentin, coupled to fluorescein. Superficially similar to Figure 6.16, this picture shows that intermediate filament proteins are flexible and form networks.
LM, ultraviolet fluorescence, anti-vimentin stain, ×440

Higher organisms contain various cell types of different specialities. What a cell does is determined by the activity of genes within its nucleus. Although each nucleus effectively contains a huge library of information and blueprints, in any particular cell only a few volumes of that library will be acting as instructions.

The way in which cells specialise may be biochemical – the production of a particular metabolite, for example. Or it may involve considerable structural modification; the development of networks of endoplasmic reticulum and the production of chloroplasts are examples already illustrated in this chapter. Cells may specialise in an ability – to sense gravity, for example, or the colour and intensity of light. Whatever their function, they act with other cells to produce a working tissue, such as a liver, or a leaf. It is this ability to act in a coordinated fashion which distinguishes the cells of highly evolved creatures from those of more simple groups.

6.18 Muscle fibres consist of a highly organised arrangement of two types of fibrous protein which slide over one another during muscle contraction. This transmission electron micrograph shows a cross-section of the fibres in the flight muscle of the fly, *Bombylius major*. The black circles arranged in a hexagonal pattern are sectioned myosin filaments. Each one is surrounded by six black dots, which are sections through actin filaments. The two filament types are linked at intervals by cross bridges – visible as the hazy grey halo around the myosin filaments. Muscle contraction involves the breakdown of large amounts of ATP, and so muscle fibres are often closely associated with energy-producing mitochondria.
TEM, stained section, ×31 000

6.19 Sperm cells are specialised for swimming: they are streamlined in shape, and their cytoplasm is reduced to bare essentials. They have no endoplasmic reticulum or Golgi apparatus, and the nucleus is very densely packed with DNA. The cytoplasm contains many mitochondria, which provide the energy for movement. The cell consists of two parts: a head region, containing the nucleus, and a long tail, or flagellum, which has mitochondria wrapped round its base. Movement is accomplished by the rhythmic beating of the flagellum. This picture shows a group of sperm tails in cross-section, during their maturation in the testis of a moth. Each contains an array of microtubules – two dead centre, surrounded by a group of nine microtubule doublets and, beyond those, nine single microtubules pressed against the plasma membrane of the cell. The central pair and the outer nine microtubules are filled with a stained material and appear solid in

this species. The beating of the flagellum is accomplished by the microtubule doublets sliding past one another in a process which is fuelled by ATP. The large granular body within the plasma membrane, at bottom left of each flagellum, is called a 'dense fibre'; its function is unknown. Also unknown is the reason why, in this species, the outer surface of the plasma membrane is decorated with striated projections.
TEM, stained section, ×83 000

6.20 The surface of a cell is usually more or less smooth, but in certain situations cells develop bristle-like extensions called brush borders. These cells are found in tissues where rapid absorption is taking place, such as in the surface of the intestine, and in the kidney. This picture was obtained by rapidly freezing a piece of kidney tissue, then fracturing it in a vacuum and coating the exposed broken surface with platinum. The micrograph is of the platinum replica.

It shows a brush border in the kidney epithelium. Each of the bristles – called *microvilli* – contains bundles of actin filaments. The effect of the brush border is to increase the surface area of the cell about 25 times. The large objects, top right and bottom left, are surface views of cell organelles.
TEM, freeze-fracture replica, ×17 500

6.21 A second type of bristle-like extension to the cell surface is shown here. Each bristle is called a *cilium*, and it contains a bundle of parallel microtubules. The microtubules grow out from 'basal bodies' – seen here in a row lying just inside the cytoplasm, and associated with a dark-staining granular material. Cilia perform a beating movement by sliding adjacent microtubules over one another. The function of the beating motion varies with the cell type. Some protozoa are covered in cilia, and use them to propel themselves through water, or to set up currents that attract food particles. The surface of the

respiratory tract in animals is also ciliated; here the beating of cilia sweeps layers of mucus, dust and dead cells up towards the mouth. Cilia also help to sweep egg cells along the oviduct. Ciliary movement requires energy, and this is supplied by mitochondria, several of which can be seen at bottom left of this picture.
TEM, stained section, ×25 000

6.22 The liver is the organ of the body which processes nutrients from the digestive tract and either stores them or makes them available for use elsewhere. In this micrograph of a liver cell from the slender salamander, *Batrachoseps attenuatus*, the cytoplasm can be seen to contain large numbers of dark-staining crystals. These are made of protein, and have formed within the lumen of the rough endoplasmic reticulum. They probably represent reserve storage material. The cell nucleus, with its nucleolus and peripheral chromatin, is bottom right.
TEM, stained section, ×6250

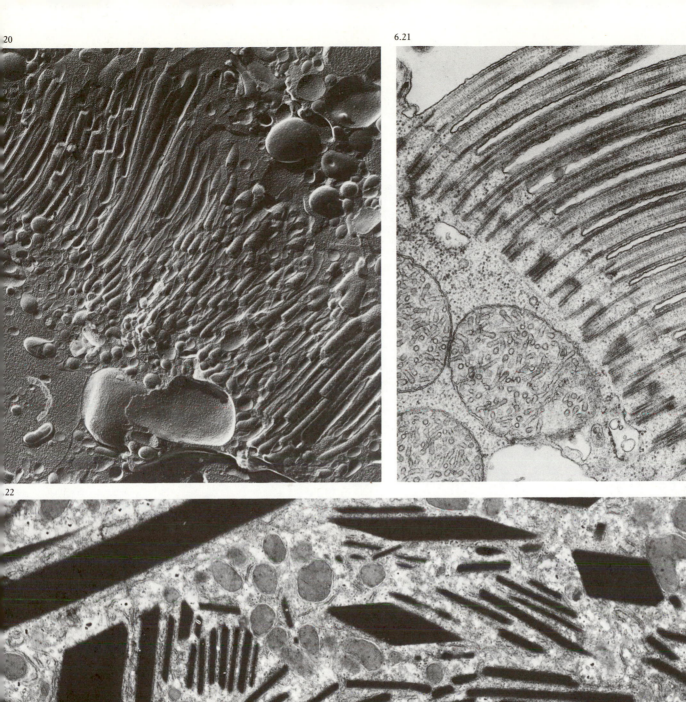

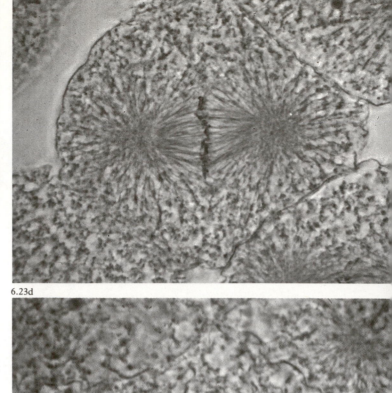

6.23c

6.23d

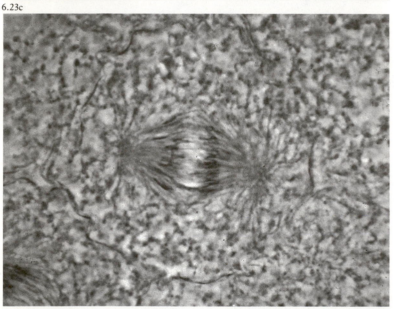

6.23e

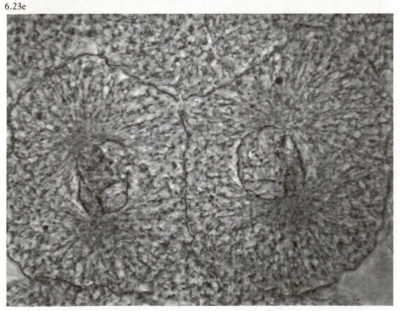

MITOSIS

The main genetic material of a cell is contained within the nucleus, where it exists in duplicate as paired chromosomes. The cell division cycle consists of two stages: interphase, when all the nuclear DNA is carefully replicated to produce the duplicate of itself; and mitosis, during which the two pairs of chromosomes are precisely distributed by a mechanical process. The result of this cycle is that both the new cell formed by cell division contain exactly the same genetic information as the starting cell. Mitosis involves an elaborate series of movements on the part of the chromosomes, mediated by a structure called the mitotic spindle.

6.23 This series of light micrographs illustrates the main phases of mitosis in an animal cell. The whole process usually takes from one to a few hours. Prophase, the beginning of mitosis, is characterised by the condensation of the chromosomes and their appearance as a tangle of dark-stained material in the nucleus (6.23a). Next, the nuclear envelope dissolves, and the mitotic

CHAPTER 7

INORGANIC WORLD

IN the modern world we are surrounded by the results of man's endeavours to improve the raw materials from which he fashions his utensils and playthings. Microscopy has played a preeminent role in the study of inorganic materials and the quest to exploit their physical characteristics. It is difficult to imagine a modern aircraft, capable of carrying hundreds of passengers, without the use of strong light alloys for the structure and tough heat-resistant alloys for the jet engines. It is equally difficult to imagine the development of the electronic chip without the availability of pure silicon crystals. These materials are only manufactured today because of the years of background study by scientists using microscopes.

The subjects of this chapter are inanimate; they are incapable of reproduction. They include metals, alloys, minerals and ceramics. Although they are all different types of materials, their study converges at the atomic level. Solids are characterised by ordered atomic structures regardless of whether the atoms are identical, as in a pure metal, or a collection of different atoms, as in an alloy or compound. The atomic arrangement or crystallography is a characteristic of a particular material. The majority of atoms within a material will be positioned in a similar manner and will be surrounded by an identical arrangement and spacing of neighbouring atoms. Crystallographers call this identical, repeating arrangement the unit cell.

Only fourteen types of unit cell can exist. This was proved by the French mathematician Auguste Bravais in 1848, and hence they are known as the Bravais lattices.

They comprise seven basic shapes and seven slight variations. The names of the seven basic shapes, which are all parallelepipeds, are: cubic, tetragonal, hexagonal, trigonal, orthorhombic, monoclinic and triclinic. The number of variations depends on the type. Cubic, for example, has three: *primitive*, *body-centred* and *face-centred*.

As long as the ordered structure prevails, a material is solid. When it becomes liquid, the atomic bonds are destroyed and the structure breaks up; the atoms are released to roam randomly throughout the liquid. If the liquid is then cooled sufficiently, the ordered structure is remade and the solid state regained.

The resistance of solids to deformation results from the reluctance of atoms to allow disturbance to their neat atomic arrangements. Deformed solids are under *internal* stress as long as their atomic structures remain deformed. Such objects are likely to change shape if they are heated to allow rearrangement of their atoms. The addition of 'rogue' alloying atoms in the atomic structure can create interatomic stresses which improve a material's resistance to *external* stress and so harden and strengthen the alloy in comparison to the base metal. The most common alloy is steel, made by combining two naturally weak materials: iron and carbon. When they are alloyed, the carbon, which is added in very small quantities (typically 0.2–0.6 per cent by weight), takes up residence in the iron's cubic unit structure and profoundly affects its characteristics. Tenfold increases of strength can be brought about, vastly improving the value of the metal as an engineering material.

The constituents of an alloy, mineral, or ceramic are known by a variety of names according to the shape, crystallography, or chemical composition that they possess. *Phase* is the most common of these names and a constituent requires a particular combination of chemistry and crystallography before it can be classed as a particular phase. This may only define it partly; an additional description by shape might be added if a variety of shapes are commonly observed.

The secrets of atomic structure can only just be observed in the transmission electron microscope. At the highest magnifications available, individual atoms can be resolved and their relationship to other atoms studied. But in order to completely characterise an atomic structure, X-ray techniques are also employed. The neat lines of atoms bend X-rays in regular patterns which can be interpreted by the microscopist to reveal not only the type and shape of the atomic arrangement but also its dimensions.

The microscope samples used in this chapter differ vastly from those encountered so far. The delicate sections of the life sciences are supplanted by the more robust but equally elegant preparation techniques of the materials scientist. Abrasives and fine polishing diamonds are used to work the specimens to mirror-quality finishes, or to grind rock samples to wafer-thin transparency. The staining techniques of the life scientist are replaced by etching with chemicals or, in the case of pure ceramics, etching by subjection to temperatures in excess of 1500 degrees centigrade.

Other specimen preparation and decoration techniques are unique to materials science. The use of reflected light enables thin coatings to be set down onto specimen surfaces to decorate phases in metals so that microstructural shapes and patterns can be discerned. Variations of hardness in alloying constituents are exploited in 'relief polishing', which causes hard phases to stand out above the surrounding matrix when prolonged diamond polishing is employed.

The use of reflected light prevails in metals microscopy in order to overcome the inherent opacity of all but the thinnest specimens. Only in petrology – the study of rocks – do microscope techniques bear a resemblance to those of the life sciences, for the petrologist employs transmitted light illumination in a majority of his studies.

7.1 The inner, atomic structure of a material controls the outward appearance of its crystalline form. Although the outer surfaces of a crystal may sometimes look complex, there will always be an exact relationship between the angles and shapes of those faces. The common salt (sodium chloride) shown in the scanning electron micrograph has a cubic unit cell and the faces of the crystal cluster are indeed arranged at right angles to each other. However, the cubic unit cell can also form crystals with the shape of octahedrons or rhombic dodecahedrons. The seven simple Bravais cells can generate 32 crystal forms and further variations increase the number to over 50. The crystalline form gives a vital clue to the identity of a material and the microscopist who studies rocks and minerals regularly will be familiar with the majority of the variations that occur.
SEM, ×400

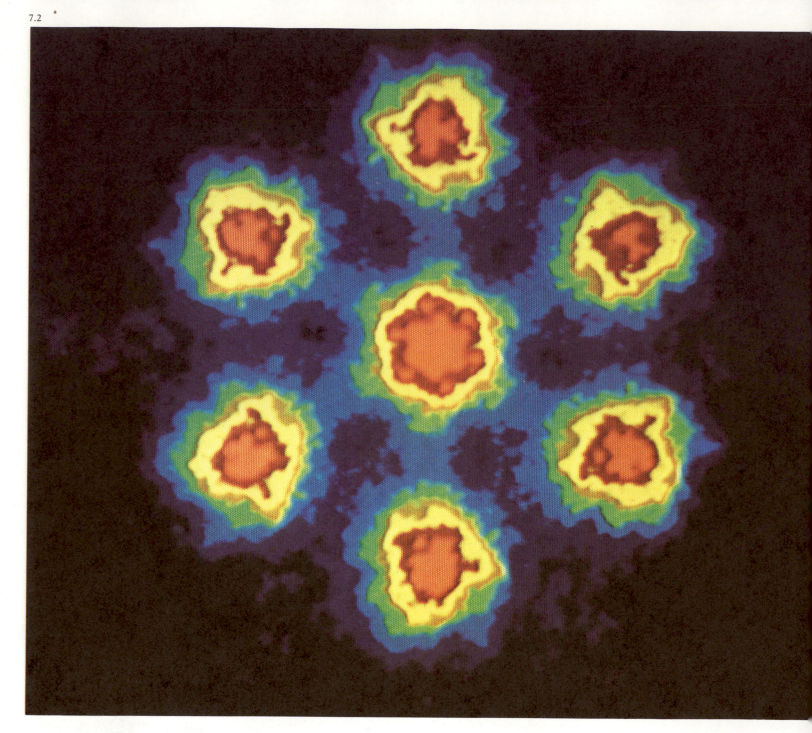

ATOMS

The smallest unit of matter that can be imaged by microscopy today is the atom. The use of high resolution electron microscopy or HREM enables the scientist to study the neat lines and rows of atoms arranged in their unit cells. The world of atomic level microscopy is bathed in hyperbole. Imaging an atom at a magnification of ×100 million is equivalent to observing from Earth the golf ball that Neil Armstrong hit on the Moon. The microscopists at the forefront of high resolution imaging are now trying to read the golf ball's number!

The HREM image is not sharp in the normal photographic sense and the image signal is frequently 'cleaned up' with the aid of a computer and complex mathematical techniques.

7.2 This picture combines scanning transmission electron microscopy (STEM) with false colour computer enhancement. The subject is a uranyl acetate microcrystal and the image shows the uranium atoms arranged in a perfect hexagonal shape around a central atom. Each atom is spaced 0.32 nanometres from its neighbour. The carbon, oxygen and hydrogen atoms, which make up the remainder of the uranyl acetate microcrystal are transparent to the electrons used as the illuminating radiation and do not show. Microscopists exploring the limits of atomic resolution have tended to use the heavy elements, such as lead, gold and uranium because they have the ability to stop electrons and therefore show up well in the electron microscope.

STEM, false colour, ×120 million

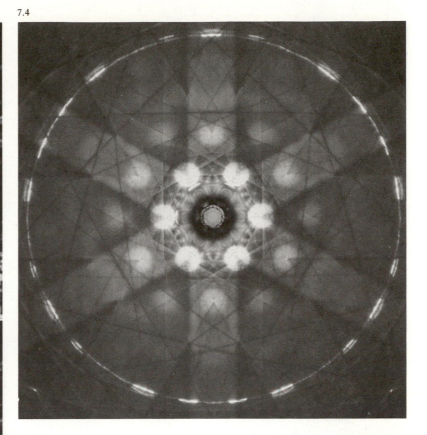

a

10Å

b

c

7.3 High resolution microscopy reveals the symmetry and order of atoms in a solid material. In nature, this symmetry is usually disturbed when a solid is subjected to stress. Distortions are produced, occurring along planes of weakness where whole blocks of atoms glide over each other in a process known as *slip*. The ability of an atomic lattice to resist slip determines the bulk strength of the material. When multiple blocks of atoms move along parallel slip planes, a process known as twinning occurs. The meeting points of slip planes often contain holes or vacancies where atoms are missing from the regular array. These vacancies are a major influence on a material's properties; they are vital, for instance, to the working of transistors and silicon chips. It is possible to observe the movements of atoms when slip occurs. These micrographs were exposed within 0.07 seconds of each other. The arrow shows the advance of a twin – the band of slanted atomic lattices – in a crystal of gold. The movement of the twin has been induced by both localised atomic forces and the bombardment of the microscope's imaging electrons. The 10 angstrom (1 nanometre) bar in the top micrograph indicates the scale. HREM, ×17 million

7.4 The pattern displayed here is a convergent beam diffraction image, also known as a Kirkuchi pattern after its Japanese inventor. These patterns are uniquely characteristic of a material and may be used to identify the smallest of particles in an alloy. The patterns are created in the transmission electron microscope by focusing a convergent beam of electrons onto the specimen and then photographing the back focal plane of the objective lens, rather than the conventional image. The back focal plane contains the 'diffraction image', which is created by the regular atomic planes within a particle affecting the passage of electrons through it. This deflects the electrons in a highly regular and ordered way which gives rise to a 'fingerprint' which can be interpreted by the microscopist. The example is from a particle of gamma phase in the refractory super-alloy 'Astralloy'. The gamma phase is responsible for Astralloy's high-temperature properties. All the lines of the image, their number, relative angles and disposition confirm that the gamma phase is present in the specimen and that the atoms are arranged in a cubic unit cell. It is not normal to quote the magnification of a diffraction image as it has no meaning. TEM, convergent beam diffraction

DISLOCATIONS & GRAIN STRUCTURES

When the scale of microscopic examination is expanded from the atomic, the simple world of ordered structures becomes more complicated. Single atomic defects like those shown in Figure 7.3 begin to collect into groups, forming 'dislocations'. Dislocations are large numbers of defects arranged in particular ways within an atomic structure and crossing many unit cells. They take a variety of forms and some reach sizes which can be observed in the optical microscope or, more rarely, with the naked eye. Dislocations can also move bodily through a solid, sometimes looping and tangling like a length of cotton shaken in a bottle of water. In the transmission electron microscope, dislocations reveal themselves as dark lines.

Many materials are made up of *grains*. Each grain has its own crystallographic orientation, although adjacent grains may have a different orientation. Grain growth can occur in a variety of ways, but the most common is from a cooling liquid. The ordered structure is preferred by atoms changing from the liquid to the solid phase because this is the most efficient way to distribute the interatomic forces. Atoms residing in the correct position in the lattice are in the lowest energy state and so they are less likely to move from these positions. The three-dimensional surfaces which enclose grains are called grain boundaries and are of considerable interest to the materials scientist. Impurities are pushed ahead of a solidifying material and are eventually trapped in the grain boundaries when adjacent grains collide. Strength is often improved by grain boundary effects, but corrosion resistance is usually reduced.

7.5 This tiny piece of a mechanically rolled sheet of a nickel-aluminium alloy displays a network of dislocations. They have piled up along the grain boundary just visible in the top left corner of the micrograph. The development of the dislocations occurred when the cold-rolled sheet was reheated or *annealed*. The density of dislocations in such a material can be as many as 500 000 million in just 1 square inch.
TEM, ×65 000

7.6 Provided the specimen has been suitably prepared and a sufficient viewing magnification is chosen, most materials will reveal a grain structure. The exceptions are the glasses, which have a special type of atomic structure. The alpha alumina ceramic in this scanning electron micrograph has a fairly typical grain structure, revealed by the network of lines. Two grain boundary impurity phases can be seen: a large one just above centre and a much smaller one at bottom right. They are beta alumina phases which contain sodium oxide, a chemical not present in the alpha alumina that makes up over 97 per cent of the material. The holes are areas of porosity, or voids, which have not been eliminated during the ceramic sintering process.
SEM, polished section, thermal etch, ×2000

7.7 This light micrograph of common brass shows that the grain structure of a metal alloy is quite similar to that of a ceramic such as alpha alumina. The specimen has been polished and then chemically etched. Grain boundaries are more readily attacked by the etchant and show up as fine lines around the more or less polygonal grains. The straight, parallel sets of lines running across the polygons are twins which show different shades of colour to the rest of the grain because of the use of polarised light.
LM, polarised light, polished and etched section, ×500

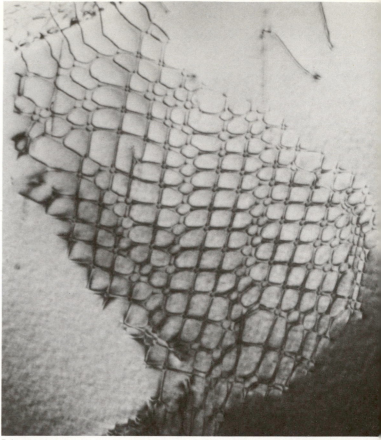

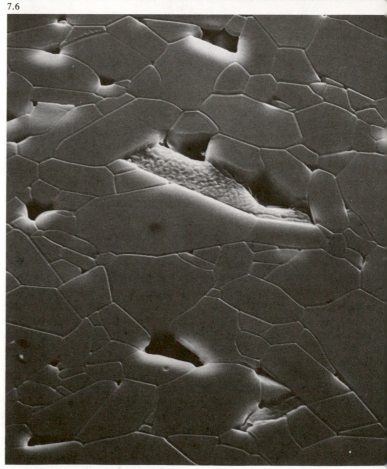

7.6

HEAT-TREATMENT STRUCTURES

Some materials possess the useful property of changing atomic structure according to how they are chemically alloyed or heat-treated. The ability to heat-treat steel has made it the most important alloy ever devised. Steel is made by the addition of small quantities of carbon to iron. Below red heat, the iron is arranged in the body-centred cubic or *ferrite* phase. The corners and centre of the cube are occupied by atoms of iron, with the carbon atoms arranged 'interstitially' between the iron. Above red heat, the centres of the cube's faces also contain iron atoms. This is the face-centred cubic or *austenite* phase. Very rapid cooling of red-hot steel will prevent the reformation of ferrite and form a brittle transformation product, *martensite*. Careful rewarming of the steel to control the breakdown of martensite will produce a steel of optimum properties – balancing the hard but brittle martensite with the tougher ferrite. Transformations like this can be observed directly if heat-treatment is carried out on specimens while they are in the microscope.

7.8 This pair of micrographs shows the transformation of austenite to martensite. The austenite (left) was held at 1070 degrees centigrade. It was then cooled to 850 degrees centigrade in 8 seconds so that it transformed into platelets of martensite (right). The picture was taken with the photo-emission electron microscope, which forms its image from the light given off when a specimen is bombarded with electrons. It can also view hot specimens.
Photo-emission electron microscope, ×870

7.9 This micrograph shows typical effects of rapid heat-treatment. The specimen is a magnesium oxide ceramic. It has been heat-treated with a laser beam, and three structural morphologies are shown. The smooth grains at bottom have a well-sintered microstructure – the powder from which the ceramic is made has compacted together at a temperature below its melting point. Above the smooth grains, the ceramic has been melted by the laser beam. The granular region consists of small crystals that have grown only a limited amount because of rapid cooling. At the top, in contrast, more prolonged melting followed by less rapid cooling has produced grains with a 'fir tree' or dendrite structure.
SEM, ×750

7.8

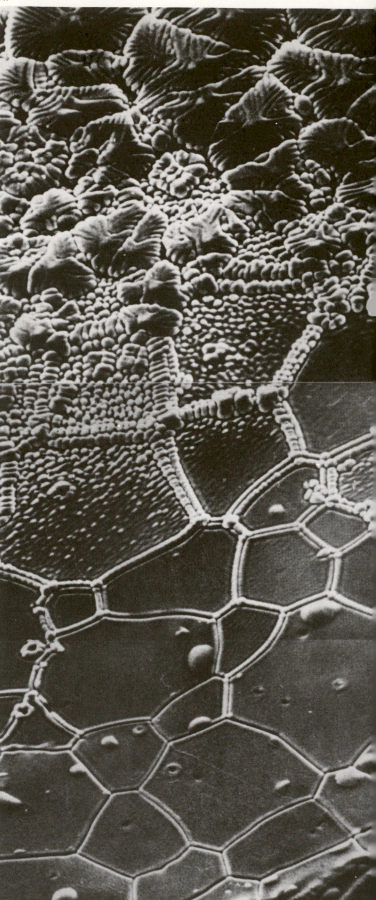

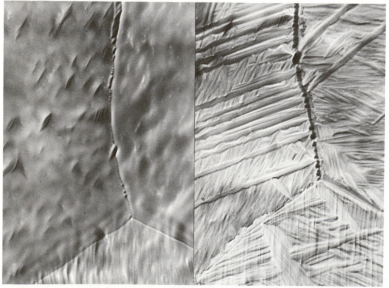

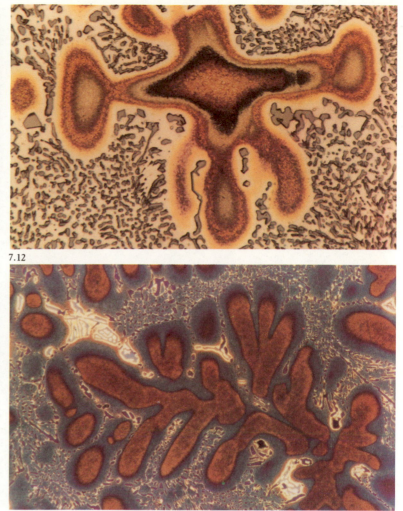

7.12

DENDRITIC STRUCTURES

The fairly regular polygons in the lower part of the previous micrograph are the simplest of a whole variety of grain morphologies. The type of grain depends on both the material itself and the way it has been made. When materials are heat-treated, changes occur in their atomic and grain structures and further changes occur on cooling. The dendritic, or tree-like, structure is typical of materials made by casting – solidified in a mold from the molten, liquid state. The process of solidification is both complex and dynamic. The atoms moving from the liquid phase to join the ordered crystal structures of the solid phase prefer to attach themselves to particular atomic planes. Solidifying atoms also emit heat and this aids the growth of 'fingers' of solid within the melt. These primary dendrites rapidly send out secondary arms at specific angles and then tertiary arms at other specific angles until tree-like structures are formed. Solidification is complete when the liquid between each 'tree' finally becomes solid. In a pure material, many dendrites commence growth simultaneously and collide with each other. The boundaries between them form the grain boundaries seen on previous pages.

7.10 These dendrites of an aluminium–titanium alloy (Al$_3$Ti) were grown under controlled conditions and then exposed by acid in order to study the angular relationships of crystal lattices and dendrite arms. Several primary dendrites have grown outwards from the left-hand edge of the picture. The secondary dendrites that grow upon them are notable for being exactly parallel to their neighbouring primaries.
SEM, acid extracted, ×100

7.11 When an impure liquid (or an alloy of several metals) is cooled, the dendrites are formed from the phase which solidifies at the highest temperature. The remaining melt, depleted in the material that formed the dendrite, will solidify into the interdendritic spacing and thicken the arms of the dendrite. The light micrograph shows a section through an inhomogeneous dendrite formed in this way in an aluminium–silicon alloy used in die-casting. Colour etching has revealed the dark brown, diamond-shaped centre of the dendrite. Around this core are several 'skins', coloured olive green, light brown, and yellow-white, which form side arms to the left and right of the central diamond and lobes above and below it. These skins are formed from increasingly depleted variants of the material that formed the core. Once the dendrite was completed, the remaining material solidified as the eutectic grey and white structure which fills the rest of the field of view. A eutectic is a material which solidifies *en masse* and with the same overall composition at a given temperature; as a result, eutectics have a characteristic 'speckled' structure.
LM, polished section, Weck colour etch, ×100

7.12 This alloy is similar to the one in Figure 7.11 except that magnesium and iron have also been added. The red-brown dendrites are enveloped by a dark blue phase. The interdendritic eutectic phase now has two variants, a predominantly white aluminium–silicon–iron eutectic and a predominantly blue silicon–aluminium eutectic.
LM, polished section, Weck colour etch, ×100

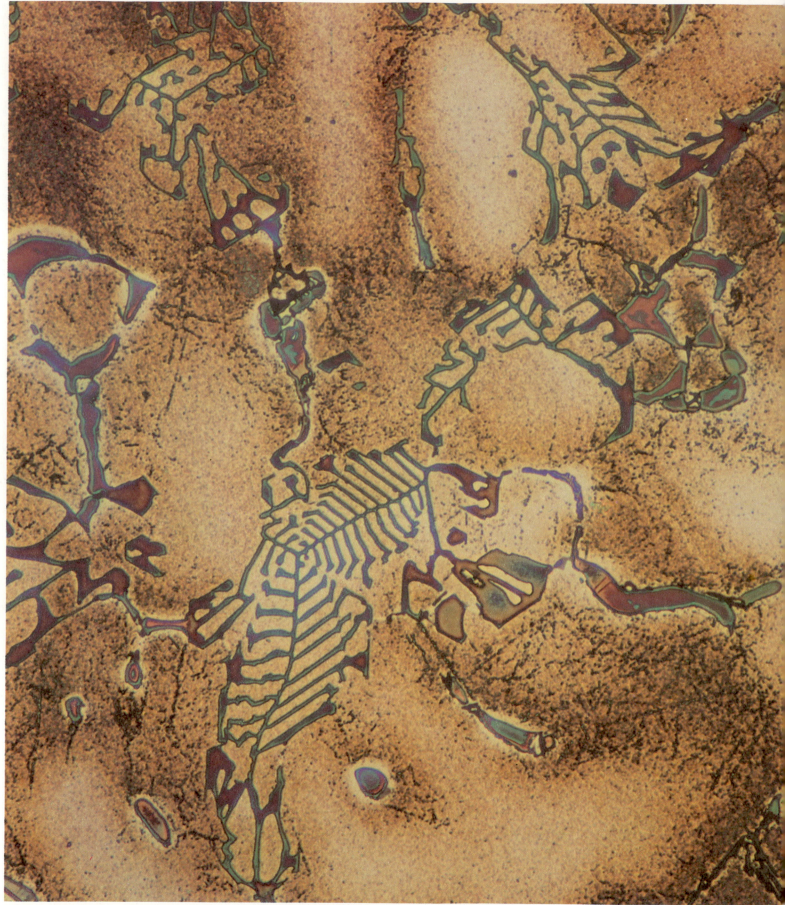

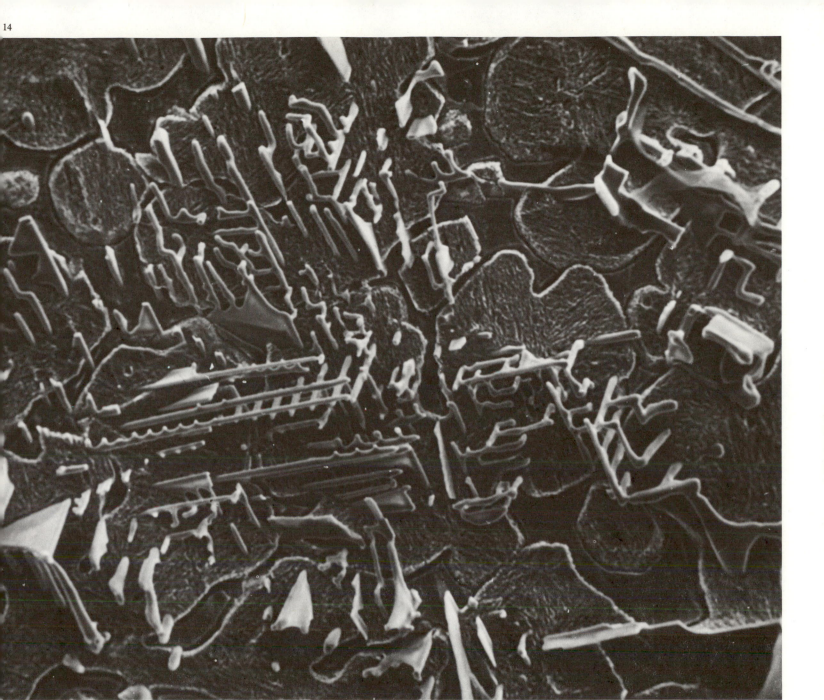

13 A large variety of aluminium alloys have been devised to meet the demands of industry. The aluminium–silicon alloy described on the previous page is a binary alloy of eutectic composition – aluminium with approximately 11 per cent silicon. In order to control the size of the dendrites, which can reduce strength if they are too large, the casting melt can be modified by the addition of small amounts of sodium metal. Although this controls the metallurgical quality of the actual casting, it is not possible to remelt the metal remaining in the ladle or any scrap from the molding process. Reducing the silicon and replacing it with copper removes the need to modify the melt with sodium and also improves the alloy strength. Other elements can also be added to improve the castability of the alloy and to effect microstructural changes. These extra alloying elements form complex and numerous phases. The aptly named 'Chinese script' structures in this light micrograph consist of dendrites of aluminium–copper–iron–manganese surrounded by the primary or host dendrites of aluminium–silicon alloy. The spaces between the primary dendrite arms are filled in places by an intermetallic chemical, copper aluminide. These are the small magenta areas with green rims. There is an overall variation of brown shading on a white background across the field of view. After polishing, the sample was immersed in a chemical colouring etchant. This solution deposited the shades of brown according to variations in the chemical composition of the sample – an effect known as 'coring'.

LM, polished section, ammonium molybdate colour etch, ×50

7.14 The complicated shapes that can develop in dendrites are well illustrated in this scanning electron micrograph. The sample is high-speed steel of the type which might be used for the cutting tool of a metal-working machine. The ladder-shaped dendrites have been exposed by prolonged acid etching and consist of metal carbides. The form of the dendrites, and the way their growth has influenced the microstructure around them, results from the use of niobium as an alloying element in the steel.

SEM, deep etch, ×2000

CRYSTAL STRUCTURES

The underlying atomic order that gives crystalline solids their symmetry is only visible in the electron microscope. But the resulting crystal shape is often large enough to be seen in the light microscope or by the naked eye. These large crystals are more in keeping with the layman's idea of what a crystal looks like.

When crystallisation takes place in a narrow space such as the gap between a microscope slide and a cover slip, illumination of the specimen by polarised light produces some of the most vibrant imagery of which microscopy is capable. Although the normal three-dimensional crystal is unable to develop in these circumstances, the underlying atomic symmetry is still present and affects the light waves. The components of white polarised light are diffracted differently so as to give rise to pure spectral colours which give these images their brilliant clarity.

Polarised light observation and photography of crystals is a favourite occupation of many amateur microscopists. For one thing, it is very easy. With the exceptions of glass and amorphous substances such as soot, almost every material crystallises. Many commonly available elements and chemicals can be made to crystallise on a microscope slide either by melting them and then allowing them to cool, or by dissolving them and then allowing a droplet of the solution to evaporate.

7.15 This very pure sulphur was allowed to solidify between a microscope slide and cover slip. Only a 'microscopic' amount of sulphur was used to fill the gap – less than 10 micrometres wide – between the two pieces of glass. The slide was heated until the sulphur melted and then allowed to cool. The cooling caused

.17

spontaneous formation of crystals, which grew out in all directions until they collided, producing irregular grain boundaries. Further cooling produced 'microfractures', which are seen as the series of fine, almost parallel dark lines. The small black dots are voids.
LM, cross-polarised light, ×120

7.16 The crystallisation of chemicals can produce dendritic structures similar to those in solidifying metals and alloys. Although the substances involved are vastly different, the feathery secondary dendrites in these crystals of the pain-killing drug Distalgesic are almost identical to those of the aluminium–titanium alloy in Figure 7.10.
LM, cross-polarised light, ×45

7.17 This scanning electron micrograph of 'fur' from an electric kettle could easily be confused with a horticultural subject. Kettle fur consists of needles of calcium sulphate which precipitate out of hard water into regular crystallographic shapes. The geologist knows calcium sulphate as anhydrite. Its monoclinic crystallographic lattice and flower-like clumps of needles are also the most common form in nature and it is from rocks bearing anhydrite that water derives its hardness. We mine approximately 10 million tons of anhydrite (and its hydrate – gypsum) per annum for building materials.
SEM, ×500

7.18–7.19 These two micrographs are both of vitamin C, a substance which shares the monoclinic crystallography of anhydrite. The scanning electron micrograph could easily be a close-up of the kettle fur in Figure 7.17, since it demonstrates a similar crystal habit. The light micrograph, on the other hand, is quite different in its appearance and perfectly illustrates the effect of constraining the crystal growth to a thin section. This could be achieved in the same way as the sulphur in Figure 7.15 or by allowing a small drop of vitamin C solution to evaporate on the microscope slide.

The cruciform 'interference figure' which appears in the centre of the 'eye' in the light micrograph is a typical polarised light feature and results from the polariser and analyser being at right angles to each other in the optical train of the microscope. The interference figure is normally observed with the eyepiece of the microscope removed; studying its movement as the sample is rotated enables the microscopist to determine which type of crystallographic group the sample belongs to. As with the sulphur in Figure 7.15, a certain amount of microfracturing has occurred. In this instance it is the stress produced as the material dried which has caused the sample to crack. The direction of the stress was at right angles to the fractures and hence was acting almost at right angles to the grain boundary which splits the picture through the middle and deflects around the 'eye'. The jagged 'mountain peaks' running down both sides of the picture are incompletely formed dendrites, which stopped growing when all the available water had evaporated.

7.18 SEM, ×80
7.19 LM, cross-polarised light, ×150

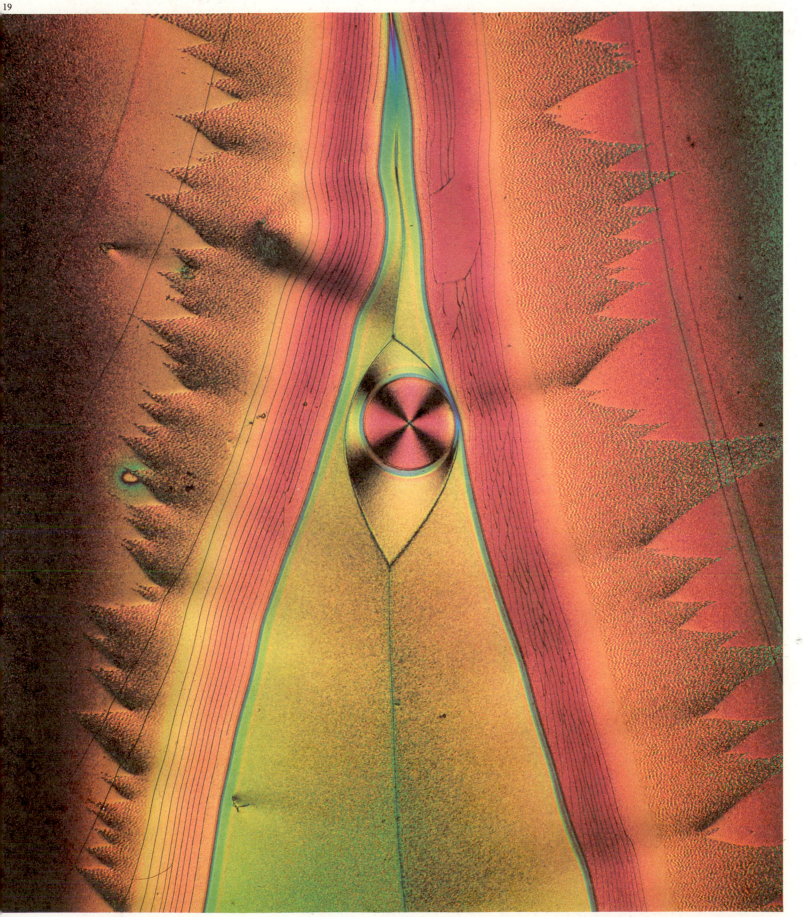

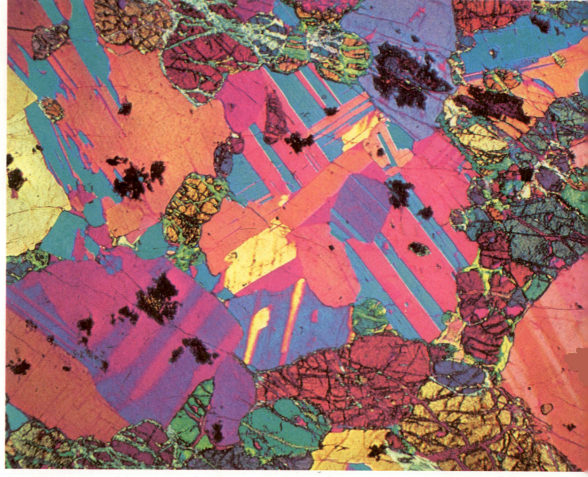

The study of rocks is a branch of the geological sciences and is known as petrology. In some ways petrological microscopy bridges the gap between life science and material science microscopy. The petrologist employs the thin section of the life scientist but shares all the concepts of crystallography, solidification and phase with the metallurgist. The petrological microscope is a specialised instrument which has sophisticated polarising facilities and an accurately rotatable stage. Specimens of rock to be examined are glued to glass microscope slides and then ground down until they are 30 micrometres thick. At this thickness, the minerals within the rock are mainly transparent, but when examined in cross-polarised light they show a variety of colours and shades (including grey). Knowledge of these colours and how they change on rotation of the sample enables the petrologist to determine which minerals are present in the specimen.

The petrologist is the historian amongst microscopists – his subjects are generally as old as the Earth. The Earth's core contains a molten mix of rock-forming chemicals called magma which is periodically ejected through the Earth's solid outer crust by volcanic processes. The solidified magma forms *igneous* rocks, the nature of which depends on the composition of the original magma and its thermal history during the process of ejection and immediately afterwards.

7.20 Gabbro is an igneous rock which contains the minerals olivine and plagioclase feldspar. The former is recognized in the microscope by its multitude of randomly oriented fractures, the latter by the multiple twinning of its lath-shaped crystals. In this polarised light micrograph, the twins are seen in a variety of colours because of the use of a sensitive 'tint plate' in the optical train of the microscope.
LM, cross-polarised light with tint plate, ×36

7.21 The upper four pictures on the right-hand page are also igneous rocks. First is a Cornish granite, which has a more complicated mineral assemblage than gabbro. Quartz, orthoclase and plagioclase feldspars, biotite and chlorite are all present. The variously shaded grey crystals are quartz; the twinned crystals are plagioclase feldspar. The mineral biotite shows as a variety of bright interference colours. The mid-grey crystals intruded with veins of a white mineral are orthoclase hosts with albite intrusions.
LM, cross-polarised light, ×7

7.22 This sample is a porphyritic basalt from Mont Dore, France. It displays a quite different texture to granite, although it shares one common mineral – plagioclase feldspar. The feldspar shows clearly as long, 'zebra-striped' twins in a very fine basaltic matrix. Large crystals in a fine-grained matrix are known as *phenocrysts*. The brightly coloured phenocrysts in the specimen are pyroxene and olivine. A feature of this type of rock is that the fine-grained matrix shares the same chemical composition as that of the phenocrysts. The actual size of the phenocrysts depends on the rate at which they were cooled out of the magma.
LM, cross-polarised light, ×7

7.23 This porphyritic rock from Aberdeen, Scotland, displays an even greater difference between its fine matrix and large phenocrysts. This is due to slow cooling during the time when the phenocrysts of clear, grey quartz and turbid, orthoclase feldspar were growing, followed by rapid solidification of the matrix as the volcanic lava was ejected onto the Earth's surface. The large twin towards upper right is a Carlsbad twin, a simpler variation of the multiple twins of plagioclase feldspar.
LM, cross-polarised light, ×7

7.24 Cyclic variations in composition and texture can occur within a single phenocryst in a process known as *crystal zoning*. The highly coloured phenocrysts in this specimen of plagioclase feldspar are all zoned, but the blue and green hexagon at lower right is the most clearly marked. The zoning has occurred because of the changing chemical composition of the molten magma as the phenocrysts grew and floated towards the Earth's surface prior to solidification. The solidification stopped further changes and locked all the variations in place.
LM, cross-polarised light, ×7

7.25–7.26 The deposition of igneous rock is followed in the 'rock cycle' by weathering. The agents of weathering are rain, wind and temperature, which combine to fragment the rock and transport it to lower ground. Materials transported from their source eventually settle as sediments and form *sedimentary rocks*. If sedimentary rocks are buried, they are subjected to heat and pressure and become *metamorphic* rocks. These can be exposed and weathered again, or they may be fully melted and then ejected once more as volcanic rocks to complete the rock cycle.

23 7.24

25 7.26

Figures 7.25 and 7.26 are
sedimentary and metamorphic
specimens respectively. The former is
'Greywacke', composed of variously
sized fragments in a generally fine
matrix. This is typical of deposited

materials which have not been
transported very far from their source
and have had little opportunity to
become smooth and rounded. The
dark, oval grain at right of centre
consists of a multitude of sutured

quartz grains and will have been
weathered from quartzite, a
metamorphic rock.

Figure 7.26 is a sample of quartzite,
which is also known as 'Arkose'. The
specimen, from Ord, Isle of Skye,

consists of more evenly sized and
rounded crystals than the Greywacke.
This is evidence of protracted
transport, which has both smoothed
and sorted the grains.
LMs, cross-polarised light, ×7

DIAGENESIS

The process of reforming rocks from weathered fragments is called *diagenesis* when it takes place close to the Earth's surface at low temperatures and pressures. The micrographs on this page show diagenesis in a specimen of sandstone which is in the process of becoming 'solid' or *lithifying*.

7.27–7.28 This pair of light micrographs shows the same thin section in unpolarised (left) and polarised light. The polarised light not only changes the colours, but makes visible structures which cannot be distinguished in the unpolarised image. The sandstone consists of quartz fragments of three types, one above the other. The green in Figure 7.27 is epoxy resin, injected into the porous rock to enable the thin section to be prepared. The upper quartz grain consists of an egg-shaped core, delineated by a brown haematite skin. Its rounded nature tells us that the particle has been shaped and polished by transportation. Around the core, quartz has deposited chemically to produce a crystal-shaped 'overgrowth' with flat faces. The fact that in the polarised light picture the grain's core and overgrowth have the same grey tone indicates that the overgrowth is in perfect alignment with the core. The middle quartz grain consists of 'sutured' subgrains aligned north–south and crossed by the birefringent accessory mineral, muscovite, which appears highly coloured in Figure 7.28. This grain has had a different history to the first grain; it is *schistose* quartz, and has been squeezed during a metamorphic rock-forming process. The bottom grain is of a third type. Although it looks more or less homogeneous in the unpolarised light picture, Figure 7.28 shows clearly that it consists of subgrains that have been joined in random crystallographic orientations, creating quite different colours in the polarised light.
LMs, unpolarised and cross-polarised light, ×150

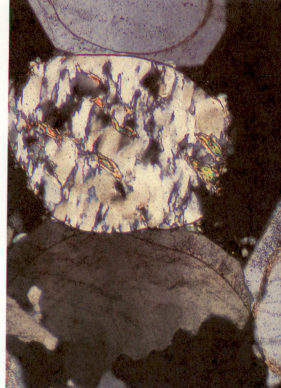

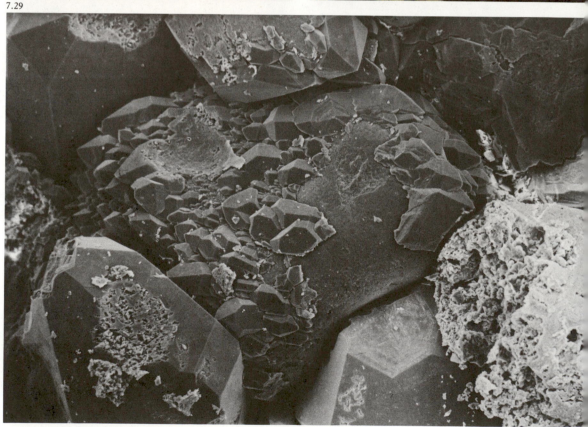

7.29 This scanning electron micrograph shows a part of the same rock from which the thin section above was made. The grain at centre is almost covered with small, angular, crystal overgrowths. This ragged texture is characteristic of partly lithified rock. Such rocks are commercially important because the partly filled voids are capable of holding oil or gas.
SEM, ×100

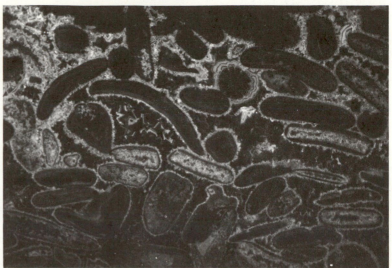

30–7.31 This sedimentary limestone ⁣emonstrates the use of ⁣thodoluminescence to reveal the ⁣ ⁣nermost details of a rock structure. ⁣he rock consists mainly of sea-shell ⁣agments compacted together and ⁣en cemented with calcite. Figure ⁣30 is a light micrograph made in ⁣ane-polarised light. This reveals very ⁣tle substructure in between the well-⁣fined shell fragments. While it is in ⁣sition on the stage of a special type ⁣ light microscope, the sample can be surrounded by a vacuum chamber and bombarded with high-energy electrons. This causes the calcite cement to luminesce, as in Figure 7.31. The sample absorbs short wavelength radiation (electrons) and emits longer wavelength visible light (in this instance orange). The additional detail produced is particularly evident in the centre of the picture where the fine layers of cement can be resolved. The wheel-shaped sediment is the spine of an Echinoid, which suggests that the rocks are from the Mesozoic or Cenozoic eras. This makes them up to 200 million years old, which is very young in geological timescales.

7.30: LM, plane-polarised light, magnification unknown
7.31: LM, cathodoluminescence, magnification unknown

7.32 The geologist frequently deals with vegetable matter that has been incorporated into the structure of rocks. Typical of this process is the petrified wood shown here. The cellular structure of the wood (brown) has been invaded and replaced by silica, which has deposited from solution to form a rock replica of the original wood. Other minerals surround the wood, the bright red of agate being the most conspicuous.

LM, magnification unknown

FERROUS METALS

Metals are the most valuable asset of modern society apart from food and water. The fact that two ages of human development, the Bronze and Iron Ages, are named after metals is a testament to their importance. The metals are broadly split into those containing iron – the *ferrous* metals – and those that contain different materials – the *non-ferrous* metals. Steels and cast irons represent the bulk of the former. The addition of carbon to iron makes steel, as described on page 129. As little as 0.05 per cent carbon will begin to affect the properties of iron.

7.33 This micrograph shows a 0.87 per cent carbon steel with a *pearlitic* microstructure of thin platelets. Such a steel is about 10 times stronger than pure iron and a variety of heat-treatment processes could be employed to tailor the physical properties to those required by a design engineer. The platelets, which appear as the bright strips because they are being viewed end-on, consist of alternating sheets of iron carbide or *cementite* and pure iron or ferrite. The platelet structure is formed by the carbon being rejected from the atomic structure as the steel cools.
SEM, polished section, Nital etch, ×3000

7.34 When the carbon content of steel is less than 0.87 per cent, a mixed microstructure is produced consisting of separate areas of pearlite and ferrite. The resulting steel is more ductile, making it specially useful for those applications which require complicated shaping. This example is a 0.4 per cent carbon steel. The yellow triangular feature is a ferrite grain boundary which at an earlier stage of heat-treatment was austenite and has been 'frozen' into the microstructure during the rapid cooling. It has a herringbone or 'sawtooth' structure. The pearlite is visible as a mixture of greens and blues.
LM, Nomarski DIC, polished section, Nital etch, ×100

7.33

7.34

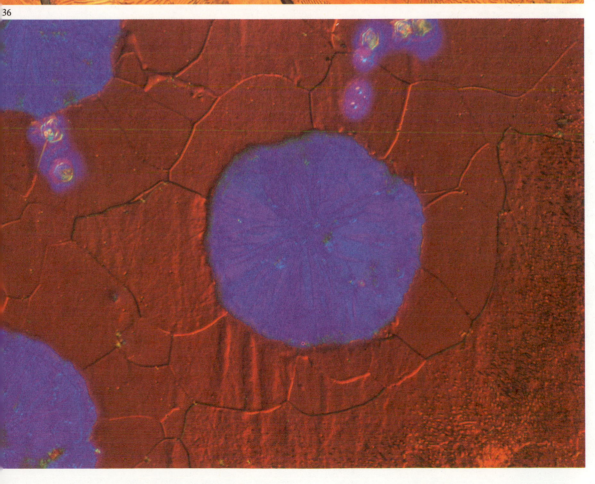

7.35 Iron can hold only so much carbon *within* its atomic structure. Once the carbon level increases beyond 4 per cent, discrete flakes or particles of carbon are formed and the steel becomes classed as *cast iron*. Cast iron is very suitable for applications requiring bulk. The presence of high quantities of carbon also makes the surface of castings hard and durable. This is a 'grey cast iron' which contains its excess carbon as graphite flakes. The flakes aid damping against vibration, a useful feature in large castings. They also make machining easier and absorb shrinkage stresses as the casting solidifies. The specimen has been coated with a thin layer of iron oxide in order to reveal the various constituents of the cast iron in different colours. The graphite flakes are deep blue, set in an orange background of pearlite. The deep orange skeleton at centre is ferrite which contains tiny particles of iron phosphide.
LM, polished section, Nital etch, ferric oxide coating, ×950

7.36 Although flake graphite confers some advantages to grey cast iron, it has the drawback of making the casting brittle and susceptible to shock forces. The shape of the graphite can be modified by alloy composition and heat-treatment. The addition of 0.04–0.06 per cent magnesium causes the graphite to form as balls. This is known as 'spheroidal cast iron' and is more ductile than grey cast iron because it lacks the sharp flakes. This specimen has also been coated with iron oxide, colouring the graphite rosettes blue against the magenta of the ferrite. Sweeping across the far right of the picture is a different form of pearlite which has a grainy appearance. The platelets have here been replaced by small spheres of cementite to form spheroidised pearlite.
LM, Nomarski DIC, polished section, Nital etch, ferric oxide coating, ×950

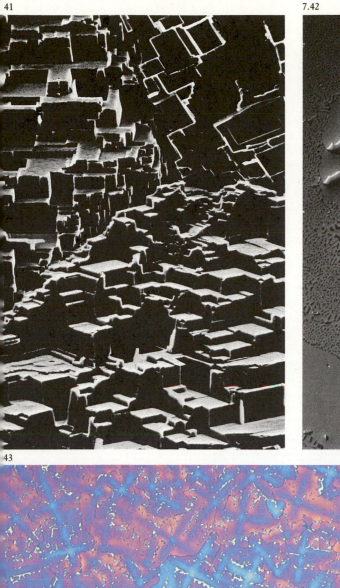

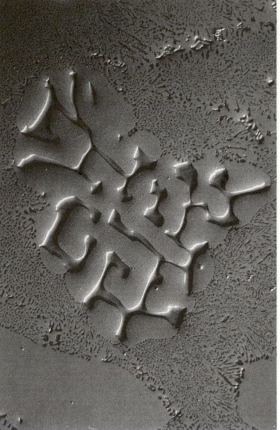

43

7.42 The addition of silicon as an alloying element improves aluminium's strength and ease of casting. In casting, the molten alloy is poured into a fixed mold so that a complex shape can be formed. The shrinkage on cooling has to be controlled by the selection of the best alloy mix. In this specimen a 'Chinese script' dendritic structure has formed, surrounded by pure aluminium. Outside this region is a fine, mottled eutectic consisting of 11 per cent silicon and 89 per cent aluminium.
LM, Nomarski DIC, polished section, ×1400

7.43 Very demanding environments exist in the centres of jet engines. The turbine blades are moving at nearly the speed of sound in the presence of corrosive, burning gases. Alloys for this type of application must be carefully chosen if the required reliability is to be achieved. Failure can be costly in both human and economic terms. Although they are expensive, combining the metals cobalt and chromium provides the basis for many 'superalloys'. Carbon, tungsten, molybdenum, titanium and tantalum are also commonly added. These elements form hard carbide particles which lock the microstructure together, preventing atomic planes from slipping over each other and conferring good high-temperature strength. The alloy designer is now confronted with a material which is almost impossible to machine, so he must also select his alloy composition for good castability so that the turbine blades can be molded instead of machined. This is why the sample shown consists of a dendritic, cast structure. The white carbides can be seen residing in the interdendritic spaces.
LM, polished section, Beraha colour etch, ×300

CHAPTER 8
INDUSTRIAL WORLD

MICROSCOPES provide a bridge between the world of scientific research and the world of industry. In research, microscopes are employed to explore and develop new or improved materials and processes. Industry takes these developments and attempts to exploit their useful properties to produce real commercial products which perform better, or cost less, than those currently available. In industry, microscopes are frequently used to confirm the quality of a product; and they play a vital role in the analysis of any failures or weaknesses. When an aircraft crashes, microscopy will be an important tool in the investigator's armoury.

A new material may have properties which are so novel that it forms the basis of a whole new technology. The best-known example is the development in the late 1950s of manufactured silicon crystals. The ability to pack thousands of individual transistors into 6-millimetre squares of silicon made possible the modern microelectronics revolution. Video games and 'smart' household appliances, pocket calculators and supercomputers all depend on the ubiquitous silicon microchip. Today, the limits of silicon-based technology have been reached and the learned journals are full of monographs about the new wonder material, gallium arsenide.

Microscopy and electronics enjoy a symbiotic relationship. Microscopes have made possible not only the development, but also the manufacture of microchips. In turn, the microchip industry has provided a spur to the development of new instruments such as the acoustic microscope, which uses sound instead of light or electrons as the imaging radiation. The acoustic microscope is beginning to play a routine role in microchip quality control and failure analysis, because of its ability to see through the chip's opaque surface to the layers underneath.

The quest of microchip manufacturers has always been to reduce the size of their devices. Throughout the 1960s, chip manufacture involved the photoreduction of the patterns of pathways and junctions that are laid down onto the silicon wafers, and this was accomplished with light microscope optics. In the 1970s, however, the patterns became so fine that they could only be made with electron microscope technology. The number of transistors on an individual microchip rose from a few thousand to several million. The electron-beam lithography equipment used to make these modern chips is indistinguishable from an electron microscope to all but the specialist.

It is not only in high-technology fields such as microelectronics that microscopes are in constant use. A great deal of modern industry is based on processes for joining materials together to make components or finished products. Depending on the nature of the materials and the type of join required, different techniques of soldering, brazing, or welding may be appropriate. Cars contain thousands of welded joints, and so do the appliances in a modern kitchen. Checking the quality of such products requires that regular samples of these welds are subjected to microscopic examination. When there are production problems, the numbers sampled are dramatically increased. Quality control of this kind is not always carried out in a leisurely academic manner; when a production line is halted, answers are needed quickly.

The analysis of failure forms a large part of the daily workload of industrial microscopy. The failure of components invariably costs money, whether as a result of repairs or because of warranty claims from irate customers, and in some cases it can cost lives. The extent of the examination depends on the consequences of the failure, but a microscope is almost always used at some stage. The reason is the nature of the atomic features of solids described in the previous chapter. The microscopic marks on a failed component are characteristic of the mode of failure and can be interpreted by the skilled investigator.

A major cause of component failure is corrosion. It has been estimated that in the United States corrosion costs industry 5 per cent of the country's gross national product. The most common method of fighting corrosion is to coat a product with a corrosion-resistant material, such as paint or enamel. Microscopy is closely linked to coating technology. For many years it was the only means of examining and measuring the quality and thickness of a coating, and although other techniques are now available, it still performs a key role.

Amongst the many and varied developments in materials science since the 1960s, one has had a special impact on the quality of life for some people. Bioengineering is the development of materials which are biocompatible and which can therefore be used to replace defective parts of our bodies without being attacked and rejected by the immune system. The replacement hip joint is the most common example. Such *prosthetic* devices need to meet a wide range of criteria. They must not only be biocompatible, they must also have the correct lubricating properties and sufficient fatigue strength to last for years.

8.1 This false colour scanning electron micrograph shows a tiny portion of a 256-kilobyte dynamic random access memory, or DRAM, microchip. The different colours clearly distinguish the layers of pathways that cover the surface of the chip. The movement of electric currents along these pathways forms the basis of the chip's operation. The three light blue pathways on this DRAM are each 3 micrometres wide and have been produced by photolithographic techniques. The latest DRAMs have pathways as narrow as 0.8 micrometres, produced by electron-beam lithography. The half-sunken pad on each of the light blue pathways is an individual transistor memory cell. A memory cell consists of a transistor plus a number of other electronic components. The sheer concentration of information on chips like this, which might typically form part of a computer's memory bank, can be grasped from the quantity of its transistors. A microchip of 256 kilobytes has 256 000 bytes; each byte consists of 8 bits, and each bit – each single unit of binary information, 1 or 0 – requires 1–2 transistors. The whole DRAM thus contains between two and four million separate transistors. A modern mainframe computer might contain 400–1000 DRAMs, together with other types of memory microchips. DRAMs are so-called because they do not store information on a permanent basis; on the contrary, their memory cells need to be refreshed every 2 milliseconds, or two-thousandths of a second. The advantage of this is that fewer transistors are needed to construct a memory microchip. They are also inexpensive, and are therefore used in large numbers in both mainframe and personal computers.
SEM, false colour, ×11 300

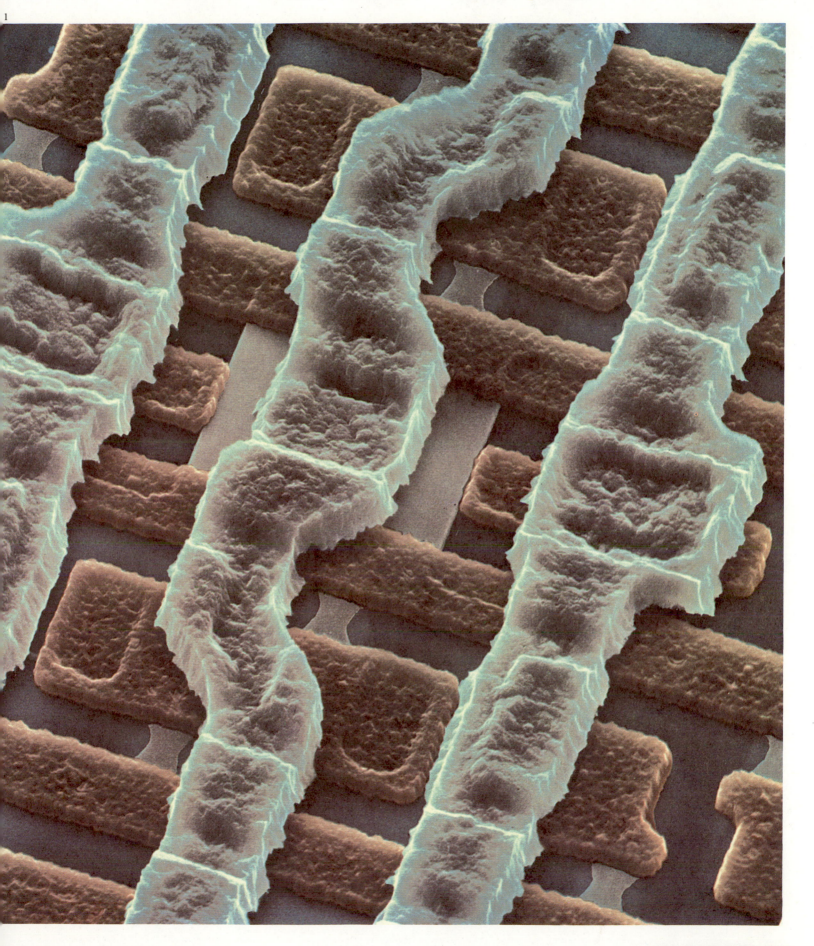

MATERIAL JOINING

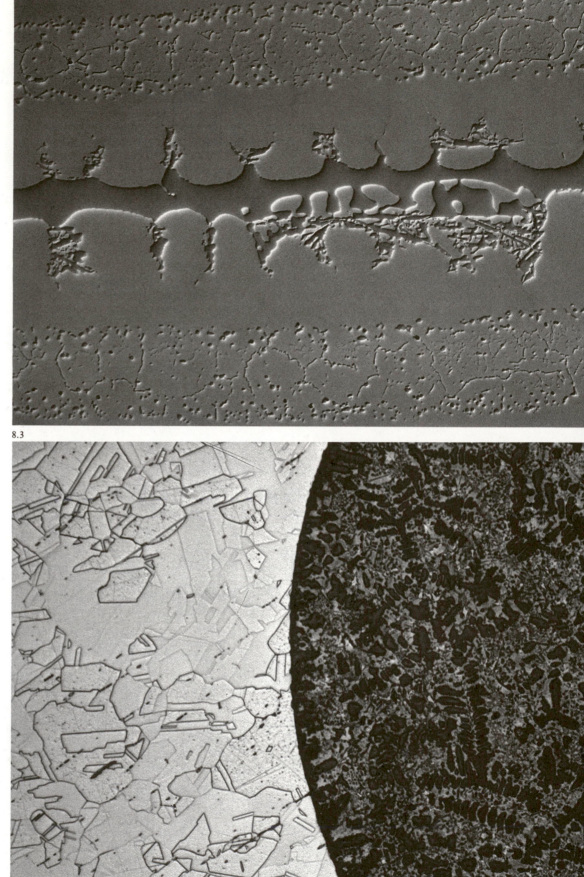

Engineers frequently join materials together in order to achieve otherwise impossible designs, or to maximise the benefits of using special combinations of materials. Metal joining techniques divide into two basic categories: *fusion* processes and *solid phase* processes. Fusion processes include soldering, brazing and arc welding; they always involve melting a metal to form the joint. Solid phase processes involve bringing materials together in such intimate contact that atoms can migrate across the gap between them and form atomic bonds. Examples of solid phase joints are friction welds and explosive welds. Resistance welds are a mixture of both solid phase and fusion techniques, made by clamping two sheets of metal between copper electrodes and passing high electrical currents between the electrodes. This process, also known as 'spot-welding', is commonly used to fabricate the body shells of motor vehicles.

8.2 Some combinations of materials or material thicknesses are difficult to weld. When welding is inappropriate but a strong joint is required, as in this aircraft part, a *braze* is often employed. Brazing alloys usually melt at around 600 degrees Centigrade and are formulated from a variety of silver, copper, zinc and gold alloys. The join line runs horizontally across the centre of the picture; the rugged features on either side of the join are dendrites projecting into the formerly molten zone. These dendrites have grown while the interface was cooling; they are the result of new alloys forming to make the joint.
LM, Nomarski DIC, polished section, etched, ×900

8.3 A frequent requirement in the electrical industry is the joining of copper to other metals. With the correct choice of brazing alloy,

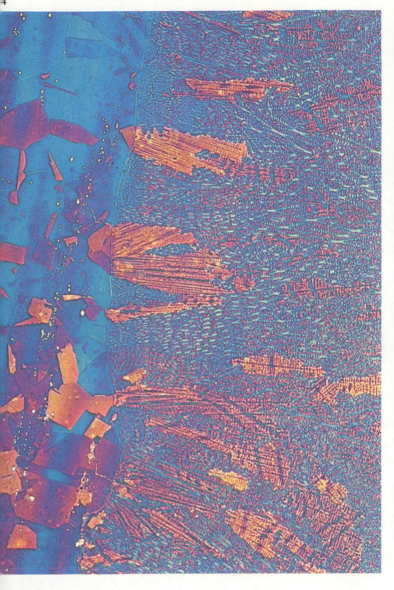

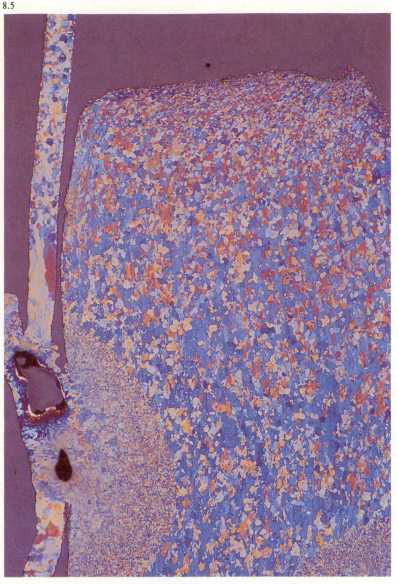

issimilar metals such as copper and ickel can be joined. This light icrograph shows the interface etween the brazing alloy and copper a such a joint. The copper is on the ft. The silver-based brazing alloy ontains well-formed dendrites, reated as it cooled and solidified. The arved interfacial area contains a dark, opper-rich phase, which formed as e silver alloy dissolved copper during e joint-making process.

M, polished section, ammonia and hydrogen eroxide etch, ×50

4 The alloys used to form the joints soldering and brazing are usually eaker, and less resistant to corrosion, an the parent metals which are being ined. These drawbacks are avoided fusion welding, where the parent etals are themselves melted so that a olten weld pool forms across the gap etween them and then solidifies to

form the weld. The melting point of the parent metals must be attained, and this temperature is almost always higher than that used in soldering or brazing. In order to prevent excessive melting, a highly localised heat source is required. A common method is to use electric arcs. And because molten metals usually react with oxygen in the atmosphere, it is necessary to blow 'shield' gases onto the weld during the fusion process. The gases protect the molten metal from oxygen. The micrograph shows the interface between a parent metal (left) and the weld fusion zone. The parent metal is a stainless steel, and has a polygonal grain structure. The weld pool has solidified into a dendritic structure which covers most of the right side of the picture. One advantage of fusion welding is that where two components made of the same metal are being

joined, the weld retains the same chemical composition – and therefore the same corrosion resistance – as the parent metal. A disadvantage of fusion welding is that it cannot be used where different parent metals have widely different melting points.

LM, polished section, Beraha etch, ×290

8.5 The microscope is frequently used to examine the quality of welded joints, although this is necessarily a destructive technique which sacrifices the weld. The example shown is a cross-section through a defective resistance weld on an electronic component. The weld is intended to join the thin sheet of iron on the left to the thicker sheet on the right. Incorrect conditions at the interface between the welding electrodes and the thin sheet have caused metal 'splattering', resulting in the two irregularly shaped holes, the largest of

which has almost severed the thin sheet. The micrograph also demonstrates other features. The area of fine-grained material radiating from the smaller hole is the heat-affected zone, where the heat from the weld interface has caused the iron to recrystallise. Recrystallisation into a finer grain size is also evident at the top of the thick sheet, indicating that it has been stamped into shape. The distortion of the crystal structure during the stamping has caused it to recrystallise at a later heat-treatment stage. The small projection in the top right part of the thicker sheet is a 'shear lip'. Shear lips are also associated with stamping processes and are responsible for the cut fingers frequently sustained when attempting to service cars or washing machines with unfinished metal edges.

LM, polished section, Klemm etch, ×70

SOLID PHASE JOINING

It is not essential to melt two surfaces together to create a welded joint. In solid phase welding, no melting takes place; the join is achieved by bringing the surfaces together in intimate atomic contact. The natural attraction between atoms creates extremely strong bonds, especially if some heat is applied.

Friction welding is the most common solid phase welding process and is used in a wide variety of applications, notably in the automobile industry. To make a friction weld, two bars of metal or plastic are rotated in contact with each other so as to generate frictional heat. During this part of the operation, one bar is slowly pushed into the other so that clean metal is 'burned' onto the weld surface. The rotation is then stopped very quickly and the bars are forced together to forge the hot, clean interface.

8.6 One advantage of solid phase welding is its ability to join dissimilar metals. This cannot be done by fusion welding if there is a large discrepancy between the melting temperatures of two materials, such as nickel and aluminium. This scanning electron micrograph shows a nickel-aluminium joint which could only have been made by microfriction welding. Small welds such as this have to be made at very high rotational speeds in order to generate sufficient surface heat. This weld would be rotated at 60 000– 100 000 revolutions per minute before being stopped in microseconds and forged together. The three protrusions at the top are the result of the forging force squeezing aluminium between the jaws of the holding chuck. These protrusions and the circular 'welding flash' would normally be machined off before the component was used in service.
SEM, ×30

8.7 It is usual when examining the microstructural quality of a friction

weld to cut and polish a cross-section. This example shows a good-quality weld between a pure iron plate (bottom) and a low-alloy steel pin – there are no voids or inclusions in the weld interface. Joining pins and tubes to plates is normally a difficult task, but it is made relatively easy by the use of friction welding. The plate is held stationary while the pin is rotated against it. Any rust or impurities on the surface of the steel plate are forced out into the weld flash, leaving the interface clean and sound. The colours of the grains have been produced by the chemical etching, which has stained them.
LM, polished section, Klemm etch, ×150

8.8 The light microscope is commonly used to study defects in welds. In this friction welded specimen the rod at the top and the plate at the bottom are separated by a horizontal tear. A failure of this kind is often the result of incorrect machine settings. If the forging pressure is applied before the rotation has stopped, for example, the freshly made weld will be torn apart. In this picture the disturbance to the grain structure from the frictional forces is shown. The vertical grain alignment of the rod at the top of the picture has been changed to a fine, brown structure which has spread out horizontally. The grain structure of the plate was already aligned horizontally and so the change is evident only by the colour change from

blue to brown. The colours are the result of the chemical etch.
LM, polished specimen, ammonium molybdate etch, ×290

8.9 Explosive welding is another type of solid phase welding and one of the most spectacular. It is carried out between two plates of metal, one of which is held stationary while the other is literally fired into it by an explosive charge. The force of impact is so great that the metals momentarily behave like liquids, creating waves at the welded interface. Such waves, complete with collapsed crests, are seen in this micrograph of steel plates coated with alternate layers of copper and nickel.
LM, ×100

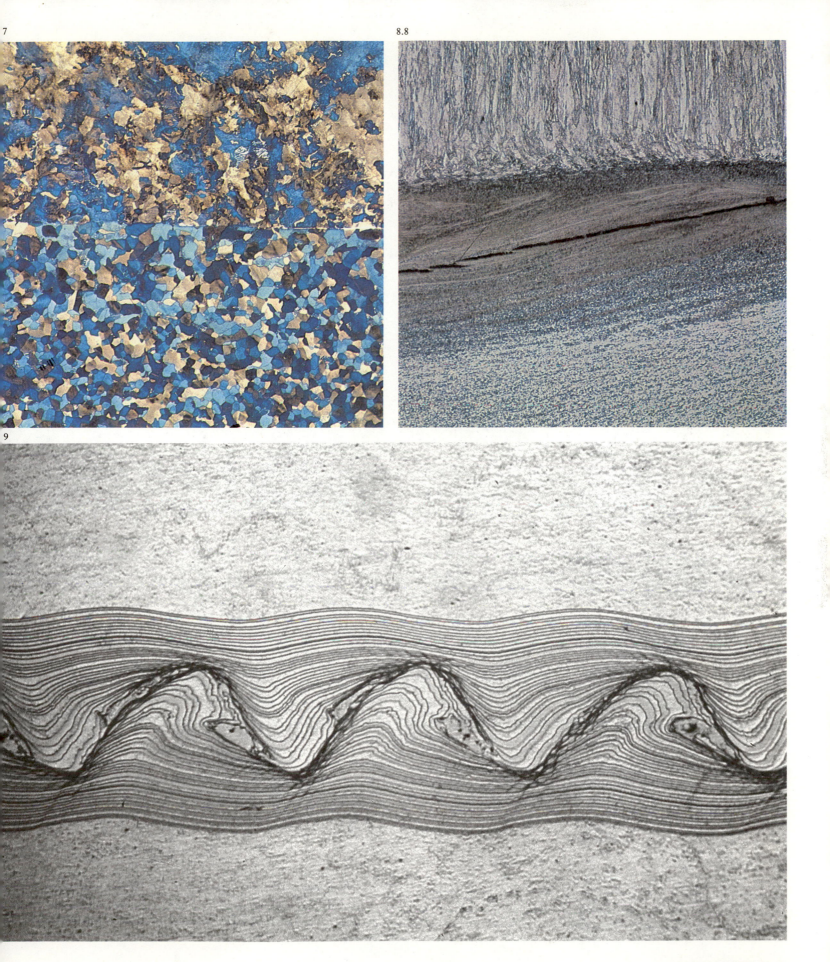

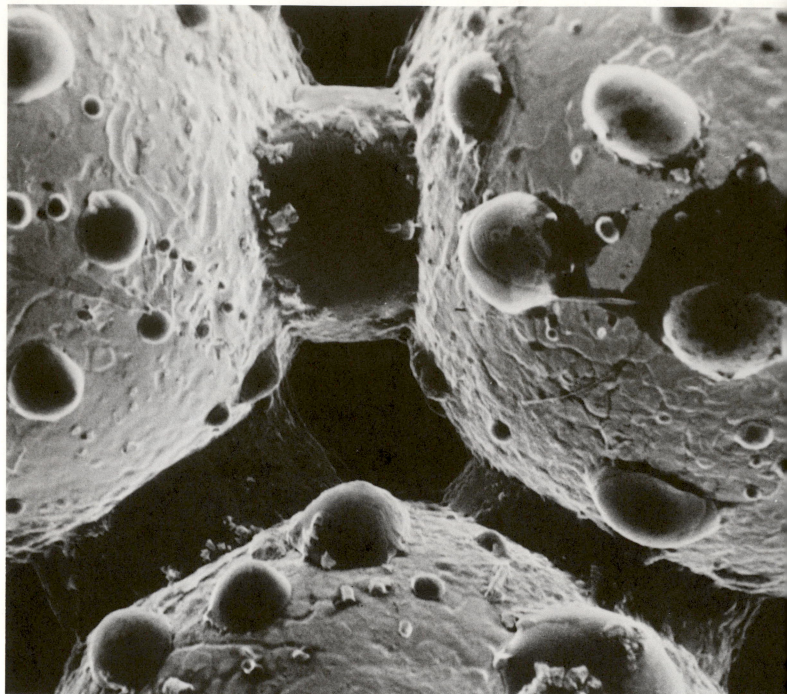

EXTREME DUTY MATERIALS

Many modern environments and duties subject materials to conditions that are enormously demanding. An example is the interior of a jet engine – an inferno in which extreme heat, stress and corrosiveness are combined. Yet we demand that the turbine blades in jet engines are manufactured to an extremely high standard, because the consequences of a failure can be catastrophic. Applications like this have led to the creation of a number of 'extreme duty materials'. In other cases, new materials have been found which have unique properties, and this has encouraged the development of new applications which can exploit their special qualities. An example is boron carbide, which is so hard that it is an equally effective – but cheaper – alternative to diamond abrasives.

8.10 The metal with the highest melting point is tungsten and it is employed for its heat resistance in such demanding applications as the exhaust cones of rocket motors. It is a difficult material to utilise, partly because its melting temperature of over 3300 degrees Centigrade means that few materials can form a crucible in which to melt it. A variety of methods have been developed to utilise the capabilities of tungsten. This scanning electron micrograph shows small spheres of tungsten bonded together by a copper 'glue'. The copper forms both the bridges between the spheres and the globules adhering to their surfaces.
SEM, ×830

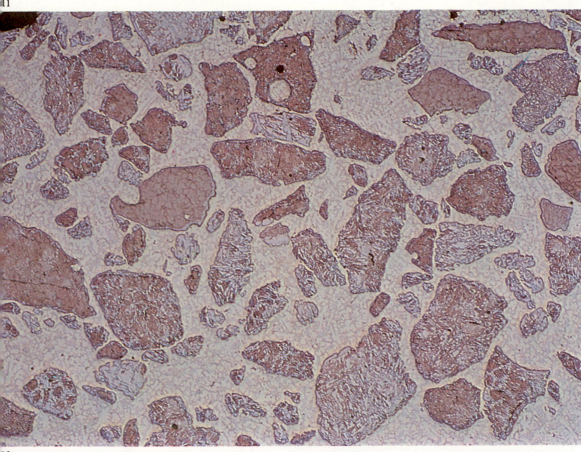

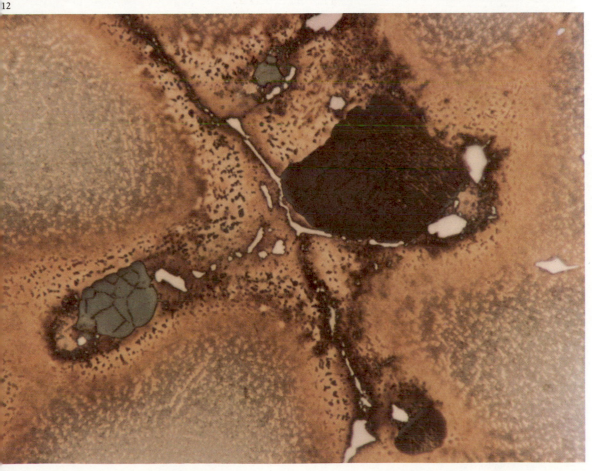

8.11 When tungsten is combined with carbon, a material of extreme hardness is produced. Tungsten carbide is capable of machining hard steels at high speeds and is very important as an industrial cutting tool. But it suffers from the drawback of being brittle. This is overcome by binding tungsten carbide particles with a copper filler. In this micrograph, the tungsten carbide particles are coloured and the copper is almost transparent. The hard, sharp, carbide particles create and maintain the cutting capability, while the copper has a very high heat conductivity and can therefore dissipate the large amounts of heat generated in a tool tip as it cuts.
LM, polished section, ×220

8.12 The turbine blades in modern jet engines are driven by corrosive gases burning at very high temperatures. The blade tips travel at almost the speed of sound and experience massive centrifugal forces. Because of these conditions, they are made from special blends of alloying elements. The metals chromium, cobalt, nickel and aluminium usually feature in the mix, together with the carbide-forming elements tungsten, molybdenum, titanium and tantalum. This specimen from a turbine blade has been etched to show the alloy's constituents. The medium and dark blue regions are a nickel–aluminium–titanium phase (the same as in the Kirkuchi pattern in Figure 7.4), while the white granules are the mixed carbides. The large grey areas are the arms of dendrites formed when the turbine blade was cast. Microscopy plays a role in both the design of alloys for turbine blades and the close scrutiny the blades undergo during manufacture.
LM, polished section, Weck's etch, ×1000

CERAMICS

Ceramics are very stable materials, which is one reason why they have often survived from antiquity. The majority are oxides of metals and the range of properties they possess is huge. Almost all are electrical insulators, have high-temperature strength and are resistant to corrosion. Ceramic products range from kitchenware to sophisticated technical ceramics used on microchips or in the tiles which cover the underside of the space shuttle. The most serious drawback of ceramics is their brittleness, but even this is being addressed by the addition of toughening chemicals.

8.13 In metal purification processes, chemicals called fluxes draw the impurities from an ore to make the pure metal and a ceramic by-product called slag. The development of this technology in ancient times helped to advance the art of metallurgy. The slag in this micrograph comes from a copper-making process that took place in Mitterberg, Austria, in 1600 BC. The section contains three circular gas bubbles in a matrix of fayalith, an iron silicate mineral. This is partially *devitrified* and consists of black granules of crystalline fayalith contained within the almost transparent 'glassy' or vitreous form.
LM, polished section, ×400

8.14 Silicon carbide has a variety of useful properties, including electrical conductivity. Although it is classed as an engineering ceramic, the geologist knows it as the mineral carborundum, the metallurgist uses it for furnace linings and electrical heating elements, and the jeweller uses it to polish gemstones. The section shown is from a *refractory brick* of the type used to line the walls of metal-smelting furnaces. The specimen has been electrolytically etched in oxalic acid to reveal its crystalline structure.
LM, magnification unknown

8.13

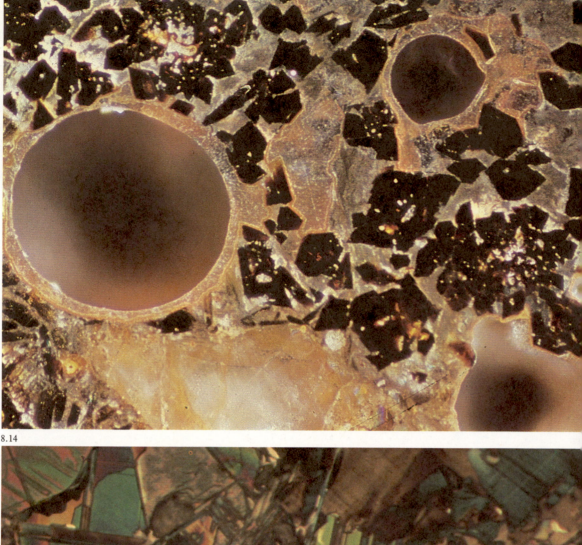

8.14

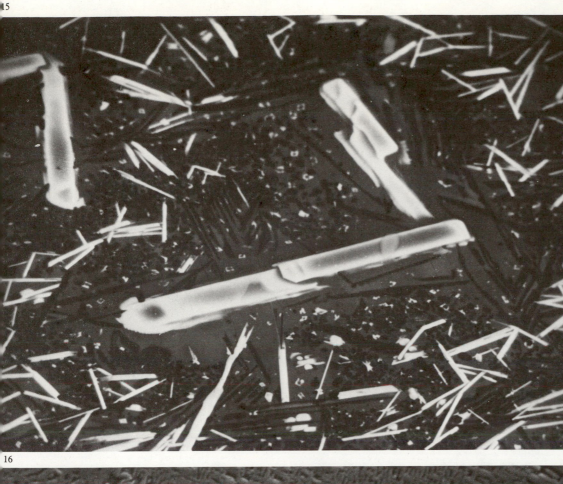

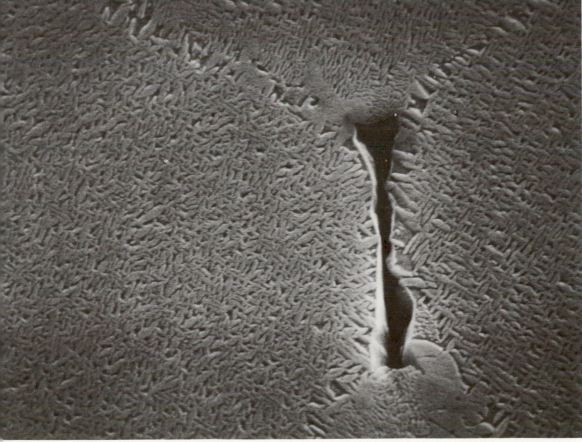

8.15 The silicon carbide in the previous picture and the silicon nitride illustrated here are both members of a group of ceramics that have become known as 'engineering ceramics', because of their use in applications such as novel types of car engine. This silicon nitride has been subjected during fabrication to a strengthening process called 'hot-pressing', in which the last remains of porosity are literally squeezed from the ceramic at high temperature. The unusual appearance of the scanning electron micrograph is due to 'back-scattered' electrons – instead of the usual secondary electrons – being used to produce the image. Back-scattered electrons are primary electrons which are reflected by the specimen – instead of being absorbed by it. They are collected by a special detector and can be used to identify particular constituents in materials. The white needles in the picture are yttrium–silicon oxides, the black needles are the mineral cristobalite, and the dark grey areas are non-crystalline, glassy phases.
SEM, back-scattered mode, polished section, ×350

8.16 One of the most exciting ceramics to emerge in the 1980s is partially stabilised zirconia, which is based on zirconium oxides. It was discovered to have the ability to change phase structure from tetragonal to monoclinic, depending on the temperature and stress to which it is subjected. The change involves an expansion of the ceramic, which fills cracks and prevents them from moving. This makes zirconia both stronger and less brittle and has earned it the nickname of 'ceramic steel'. The micrograph shows an intergranular pore with grain boundaries radiating from its top and base. Most of the picture consists of the lozenge-shaped zirconia grains.
SEM, polished section, phosphoric acid etch, ×25 000

BIOENGINEERING

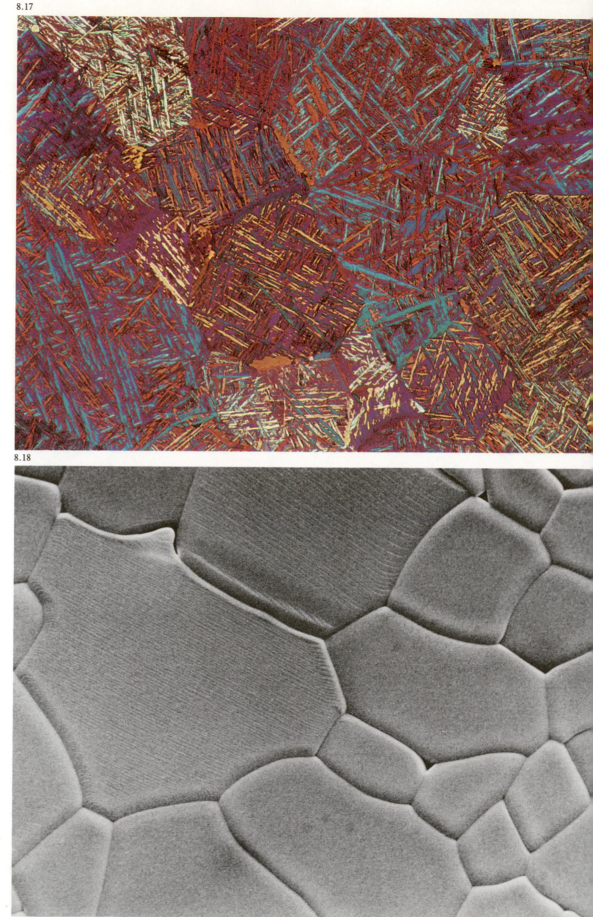

Spare-part surgery is one of the most revolutionary advances in modern medicine. It is made possible by 'bioengineered' materials which can perform biological functions. Replacing tissue, from heart valves to hip joints, with 'foreign' materials has raised many problems. The most obvious is that the implanted material must be biocompatible to prevent rejection. It must also be able to perform its designated role over long time periods.

8.17 When metal components are needed for bioengineering, titanium is one of the first choices. It has high biocompatibility combined with good strength and low weight. This specimen is a titanium alloy of the kind used in hip replacements. The titanium is cast or forged into the ball-shaped part of the joint which fits into the top of the femur. The 'basket weave' structure exhibited by the alloy is an arrangement of alpha and beta titanium phases generated by rapid quenching during heat-treatment.
LM, polarised light, polished section, ×460

8.18 The ball-shaped part of a hip joint fits into a 'cup' in the pelvis, and this alumina bone implant is used to form the cup in hip replacements. It is a 99.9 per cent pure ceramic and its minimal porosity, illustrated by the very small intergranular pores, gives it good mechanical performance.
SEM, polished section, thermal etch, ×21 000

8.18

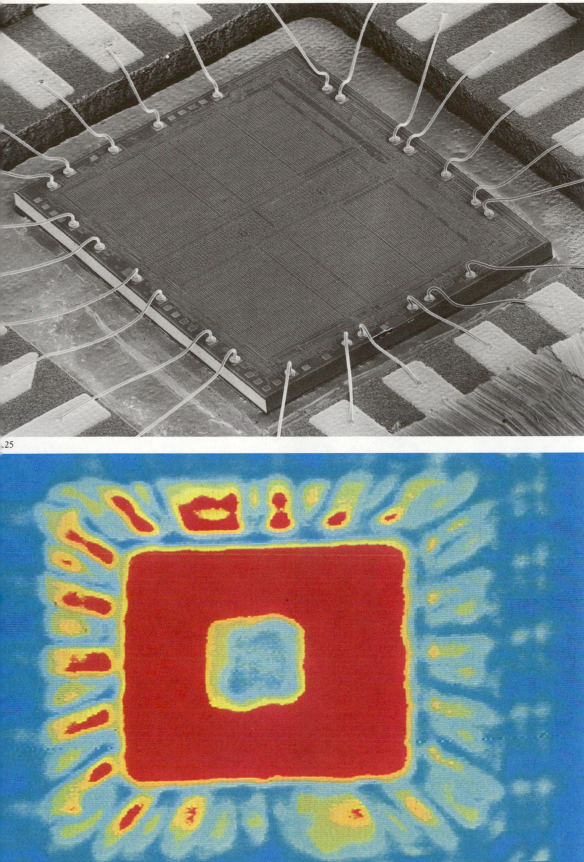

8.24 This micrograph shows the whole of the EPROM microchip that is featured in detail on the page opposite. It is seen here sitting in a well in its outer packaging. The connecting wires link the microchip's terminal pads to conductors on the packaging, where they are also welded. As more complex microchip designs have evolved, the number of connecting wires has increased. But not all the termination pads have wires welded to them. The empty pads are used for checking the microchip electrically before it is cut from the wafer. Many microchips are rejected at the final stages of manufacture because of microscopic defects on their surfaces or at junctions. This is why microchips are made in 'clean rooms' with filtered air and specially dressed operators. A simple thing like dandruff can play havoc with a batch of microchips if it escapes from beneath an operator's protective cap.
SEM, ×12

8.25 One of the factors limiting how small a microchip can be made is the heat generated by the passage of electrical current through its circuitry. In order to overcome this, silicon is frequently bonded directly to the microchip's outer packaging so that the heat can dissipate more readily. The quality of the bonding is important in such cases and a number of techniques have been tried to test it non-destructively. One of the most promising is the scanning acoustic microscope or SAM. The SAM creates an image by using sound waves, which have the valuable property of being able to penetrate beneath visually opaque surfaces. In this micrograph, the sound waves are focused beneath the surface of the microchip, seen as the red square, to show the good-quality central bond, coloured blue. The picture also shows well-bonded (blue) and poorly-bonded (red and yellow) conductors on the packaging surrounding the microchip.
Scanning acoustic micrograph, false colour, ×10

2 mm

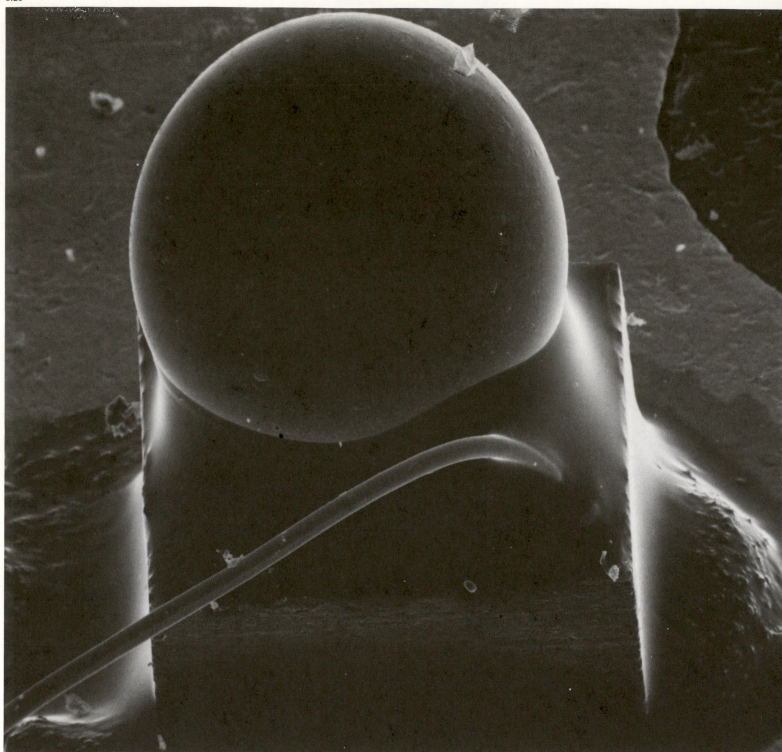

8.26 A new electronics material emerged in the early 1980s to challenge silicon's supremacy – gallium arsenide. Coated onto sapphire (aluminium oxide) substrates in so-called 'thin film' technology, it has the ability to make microchip devices operate 5 times faster than its silicon-based counterparts. This is a major advantage in supercomputers, where the speed of the devices influences the speed at which calculations can be performed. Another property of gallium arsenide is that in a particular combination with gallium-aluminium arsenide it can be used to make very small, fast-acting devices which emit light. An example of such a light-emitting device is shown in this scanning electron micrograph. The light output is of high spectral purity and it can be switched on and off 25 million times per second. Both these qualities make it ideal for digitally encoding phone conversations or other telecommunications signals which are to be transmitted down optical fibres instead of conventional copper cables. The light-generating junction within the device is situated beneath the glass bubble. The latter acts as a lens to focus the light to a fine spot, which has led to the device being termed a 'sweet spot'. The wire at the base of the bubble is the connecting wire through which the electrical input arrives to excite the light-generating junction. SEM, ×105

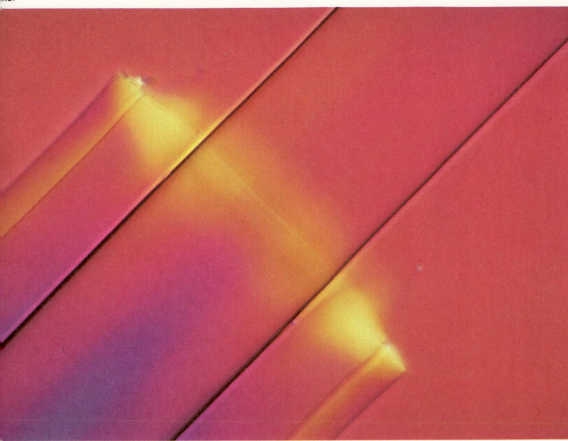

8.27 The core of an optical fibre is made of glass with a refractive index different to that of the glass which surrounds it. The effect of this is that light travels down the core of the fibre, minimising the loss of light through its surface. In this light micrograph, the core of the fibre shows a slightly deeper pink, or blue in the bottom left corner. The additional external tube at left is the fibre's protective cladding; it plays no part in the optical transmission but protects the delicate glass fibre from mechanical damage.
LM, interference contrast, ×385

8.28 Analogous to the light-*emitting* properties of gallium arsenide devices are the light-*absorbing* properties of solar cells. Solar cells absorb radiant energy from the sun and convert it into electrical energy. Most solar cells are made of single-crystal silicon, although thin coatings of silicon, gallium arsenide and cadmium sulphide are also used. This light micrograph shows the surface of a silicon crystal solar cell. The substrate consists of one type of silicon (called n-type) and onto it another type of silicon (p-type) has been deposited and formed in stepped platelets. High cost initially limited the use of arrays of solar cells to esoteric applications such as powering space satellites. As their price fell, they became more widely used – to heat swimming pools, for instance, and even to power wristwatches. Larger arrays are used in remote areas for standby power applications, and there are many prototype arrays around the world investigating their possible use to provide electricity for domestic purposes.
LM, Nomarski DIC, ×900

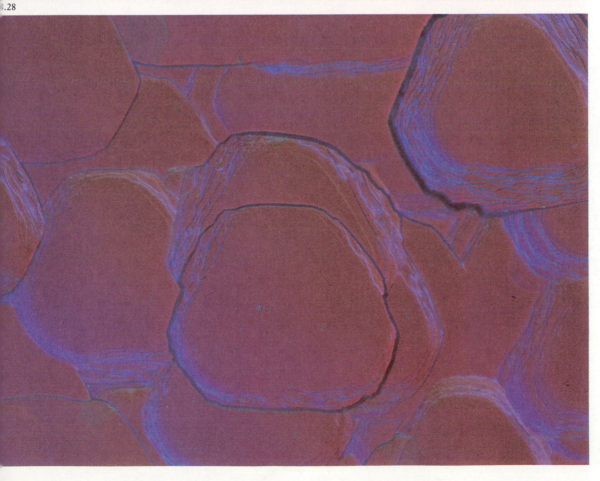

CORROSION

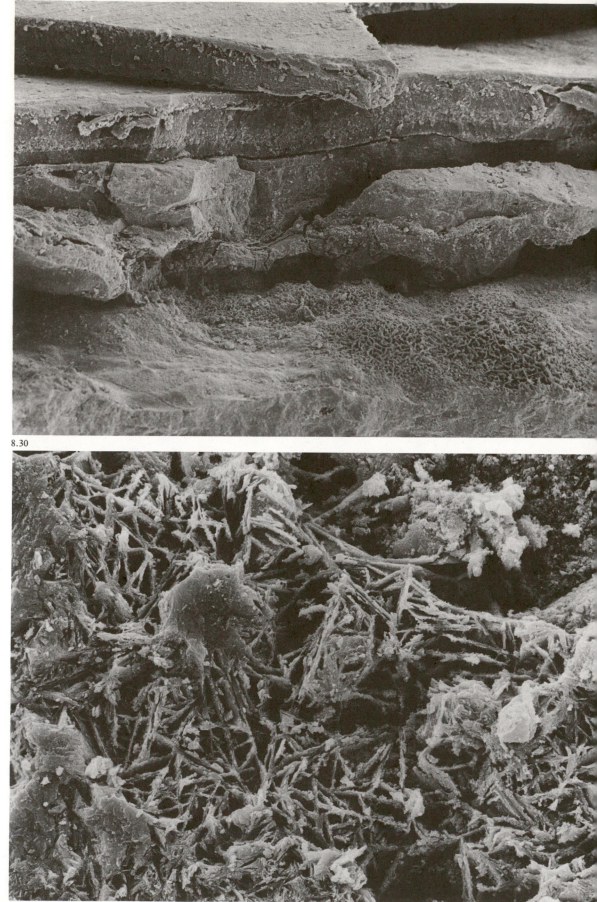

Corrosion costs billions of pounds each year. Although it takes many forms, it is usually the manifestation of an undesirable chemical reaction. In the case of common rust, it is the oxidation of steel; with old buildings, it is the attack on the stonework by pollutants such as sulphur dioxide; in chemical plants, it can be caused both by external, atmospheric sources and from the inside, as a result of the substances that are processed or stored within the tank.

Rust is the most common form of corrosion because steel is the most widely used metal. Rust forms when water and oxygen, the two corrosive agents, combine to react with iron. The oxide forms on the surface of the steel and expands until the stress this creates causes the top layer to spall off and expose fresh steel. In some instances it is the expansion itself, rather than the loss of metal, which causes the damage. For example, when an iron or steel bolt in cast concrete begins to rust, huge pressures build up which eventually shatter the concrete.

8.30

8.29 This picture shows a piece of bodywork from a 12-year-old motor car. No metal is seen. The triangular fragment at top is a flake of paint from a poor respray; beneath it are the three layers of paint sprayed on by the manufacturer. The lower third of the micrograph consists of rust. SEM, ×75

8.30 The crystalline nature of the surface of rust is shown in this detail of the specimen described above. Rust grows when moisture catalyses the reaction between iron and atmospheric oxygen. The reaction product is hydrated ferric oxide, and in this case it has grown along preferred planes to produce a crystal structure. SEM, ×825

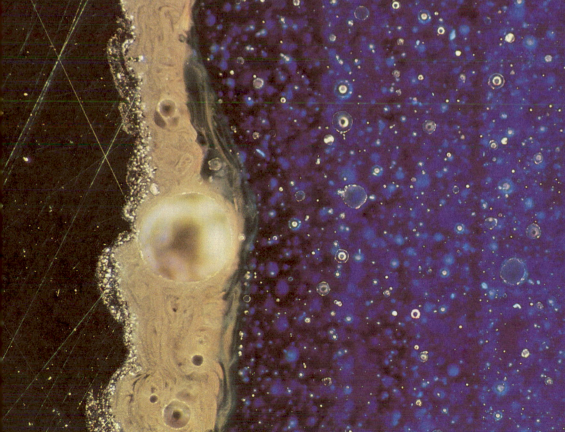

COATING

The most common defence against corrosion is to coat a component or product with a thin protective layer of a non-corroding substance. Many sophisticated coatings are available today, but the vast majority of products still rely on the basic technologies of lacquering, painting and enamelling.

8.31 This light micrograph is of a cross-section of lacquered steel. The steel is at bottom, with the colourless lacquer above it. The lacquer, which has been applied by spraying, contains four air bubbles that appear black. The good adhesion between the steel and the lacquer can be seen from the close contouring of the coating to the microscopic undulations in the surface of the steel. This is essential if the coating is to perform its function of keeping moisture away from the steel. LM, polished section, Klemm 1 etch, ×460

8.32 Enamel is one of the most durable and effective of coatings. It consists of a thin layer of glass, which has the hardness necessary to withstand the domestic abrasives frequently used for cleaning. Because glass does not directly adhere to most substances, enamelling usually involves an intermediate layer or 'interlayer' which binds chemically both to the steel, or other material from which the product is made, and to the glass. Once the interlayer has been applied, the glass is sprayed on in a powdered form called *frit*. The baking of the frit fuses the glass to the interlayer. This cross-section of an enamelled steel shows all three components: the steel in black, the cream-coloured interlayer, and the dark blue of the enamel. The circular structures in the interlayer and enamel are due to bubbles. Slightly darker vertical bands divide the enamel into five layers, which correspond to five sprayings of frit. LM, ×450

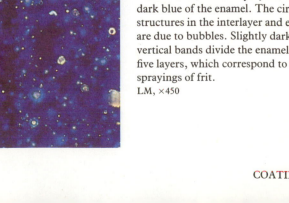

FAILURE ANALYSIS

The premature failure of devices and machines is a common occurrence. The effects of failure range from the mild inconvenience of a kitchen utensil breaking to the catastrophic loss of life when a passenger aircraft crashes. The level of post-failure analysis will vary accordingly. The broken handle of the frying pan will rate little more than an angry exclamation and a cursory inspection, but the air crash will be followed by months of intensive detective work.

Visual examination of the fragments produced by a failure plays a vital role in most investigations. The pieces of a crashed aircraft are recovered and painstakingly arranged according to their place in the original whole. The failure is reenacted in reverse and in slow motion. As clues begin to emerge, the original cause of the failure is tracked back to a particular component, which is then examined with all the microscope power available to the modern investigator.

A large part of industrial microscopy is to do with the examination of failed components. As a result, it plays an important role in the quest to give us safer transport, stronger materials, more reliable engines. The next four pages display some of the imagery of 'fractography', the science of fractured surfaces. A vital element in fractography is the fact that most breaking, tearing and rupturing processes leave identifiable and characteristic traces in their wake.

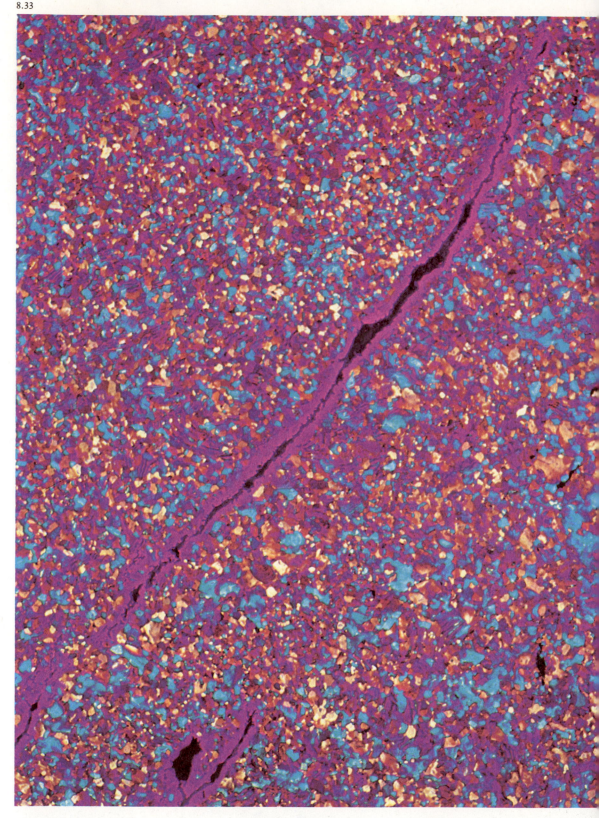

8.33 When the reason for a failure cannot be discovered by examining the surface of a fracture, a cross-section has to be prepared. Because this is a technique which destroys the component concerned, it is usually applied only after all available evidence has been obtained by other methods. This micrograph shows a section from a pressure tube made of a titanium–aluminium–vanadium alloy. The tube, of the kind used in heat exchangers for chemical processing plants, failed during pressure testing. A relatively large crack can be seen running diagonally across the picture, and a smaller crack is visible at bottom left. They were caused by intergranular tearing in the recrystallised microstructure. Although titanium has excellent corrosion resistance, it is susceptible to many common contaminants – including greasy fingers and soaps – during its heat-treatment.
LM, polarised light, polished section, ×580

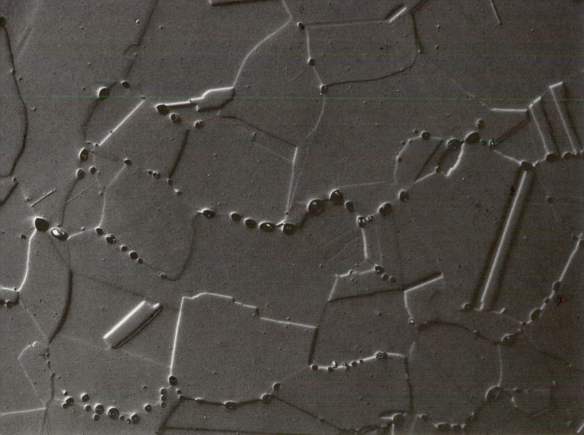

8.34 This micrograph reveals serious weakening in a brazed joint in an aircraft component. The brazing makes a 'T' joint between two pieces of nickel alloy, one forming the thick horizontal band across the top and the other the thin vertical strip that divides the lower half of the picture. The brazing alloy forms the two curving 'fillets', slightly darker grey in colour, which border the black background areas. The weakening of the joint is revealed by the small black globules within the fillets. They are indicative of excessive porosity, which could cause the joint – and hence the whole component – to fail.

LM, Nomarski DIC, polished section, etched, ×220

8.35 The fault represented by the black 'beads' in this light micrograph led to a serious explosion at a power station in Australia in the 1960s. The beads are holes in the microstructure of copper, and they are formed if copper containing copper oxides is inadvertently brazed in a furnace using hydrogen as a shield gas. The hydrogen reacts chemically with the copper oxide to produce water molecules and these turn to steam at the brazing temperature and literally blow holes in the microstructure of the copper. The holes are formed along grain boundaries and seriously weaken the copper. It was the use of this type of copper and furnace atmosphere that caused the Australian accident. The electrical fuses, which were supposed to protect the installation, contained brazed connecting tags which had been weakened. These fell off and caused so much arcing on the outside of the fuse that the gases created exploded. Fuses are intended to withstand arcing internally but the failure of the external tag rendered them useless.

LM, Nomarski DIC, polished section, etched, ×220

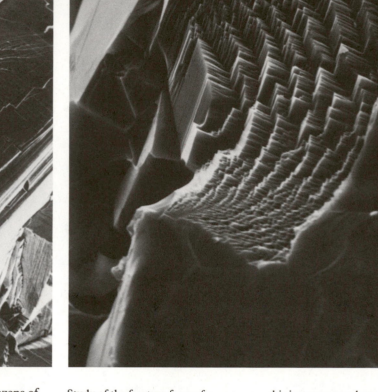

When a material is undergoing rapid, catastrophic failure, the fracture surfaces that are created are the result of highly organised effects. Even the smashing of a dropped milk bottle produces a mass of evidence for the trained microscopist. A crack moving away from the source of a failure travels quite slowly at first but accelerates rapidly to approximately half the speed of sound for the material involved – usually about 1500 metres per second. As it accelerates the energy driving the crack is dispersed in characteristic ways: first when the crack starts to weave up and down instead of travelling straight, then when it bifurcates to form two cracks. The bifurcation may then be repeated over and over again until the component

fragments – hence the dozens of pieces created when the milk bottle falls on the doorstep.

Brittle fracturing, as this process is known, occurs most commonly in ceramics, glasses and hard materials. It can also occur in steels and alloys, but in these materials it is usually preceded by a more leisurely fracturing which is accompanied by deformation of the metal. An example of this phenomenon is when a paper-clip is repeatedly bent until it breaks in two. The bending damages the metal by hardening it – 'work hardening' – until it fails in a brittle manner and breaks in two.

8.36 This scanning electron micrograph shows the fracture surface of a cobalt–chromium–molybdenum gas turbine blade from a jet engine.

Study of the fracture faces of a specimen like this enables the scientist to trace a crack to its origin. The dark grey surface in the picture contains 'river markings', so called because they join like river tributaries in the direction in which the crack is propagating. Thus the crack in this specimen has moved from right to left. The lines on the oblique surface in the top right corner are the wave-like formations created when a crack weaves up and down, as described above.
SEM, ×200

8.37 The effects of a material's atomic structure can be evident even in failure. In this specimen of the ceramic alpha alumina, a crack moving from right to left has been driven downwards. As it moved, it followed the planes of lowest energy – those which fracture most easily – in the ceramic's atomic lattice structure. In

this instance, two low-energy planes lie at right angles to each other, so that the crack has propagated downwards in a series of steps. The 'steps' stop abruptly at left where the crack plunges sharply downwards at a grain boundary.
SEM, ×2700

8.38 It is not only metals and specialist ceramics whose fracturing is studied microscopically. The 'face' in the micrograph opposite is a fracture feature in a piece of the common plastic material, polystyrene. Polystyrene is prone to reduction in the atomic weight of its long chain molecules. A year's exposure to desert sunshine is sufficient to halve its molecular weight, with an attendant loss of strength, due to photooxidation – a combined attack of ultraviolet light and atmospheric oxygen.
SEM, ×8250

QUANTITATIVE MICROSCOPY

As engineers have exploited materials more widely, they have tended to work closer to the limits of the materials' capabilities. In order to do so safely, both mechanical testing and microscopy have had to ensure that the properties of a given material are well understood and predictable. The desire for more precise information has led to a significant increase in the use of *quantitative* microscopy, where actual values are measured for observed properties and microstructures. Quantitative microscopy takes many forms, from the simple measurement of a material's hardness to the detailed mapping in a scanning electron microscope of its constituent elements. Elemental mapping relies on the X-rays emitted when the electron beam hits a specimen; each element emits X-rays of a different wavelength, providing a chemical signature.

8.39 The most common test in quantitative microscopy is the microhardness measurement of a material. A pyramid-shaped diamond, with the point facing downwards, is glued to a special, movable, front element of a microscope's objective lens. Viewing the specimen through the microscope, the investigator can see well enough around the diamond to select in a pair of cross-hairs the exact part of the specimen he wishes to test. A known weight then drives the front of the objective onto the material, where the diamond produces a tiny indent, a few micrometres across. Measurement of the indent gives a precise value of the material's hardness at that spot. The picture shows a series of microhardness indents in a metal surface.

LM, interference contrast, polished section, ×480

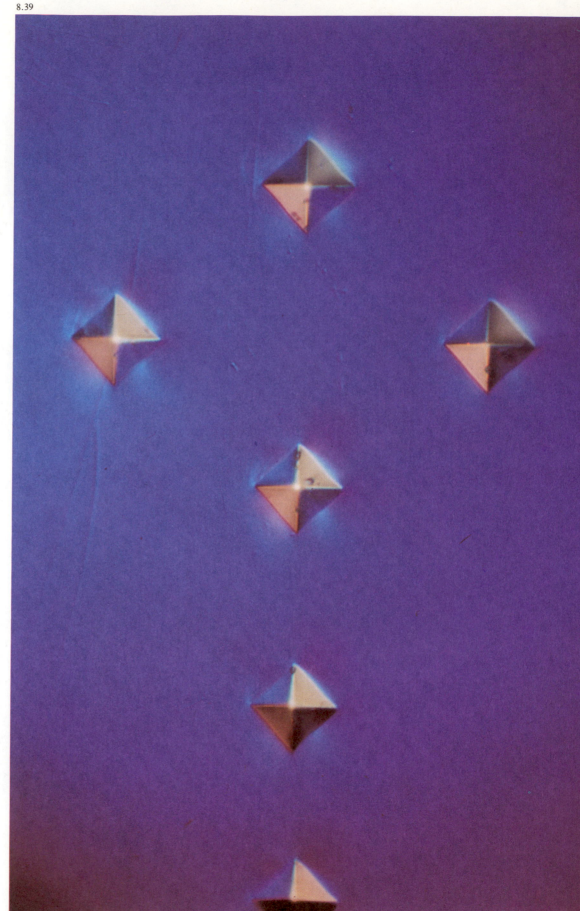

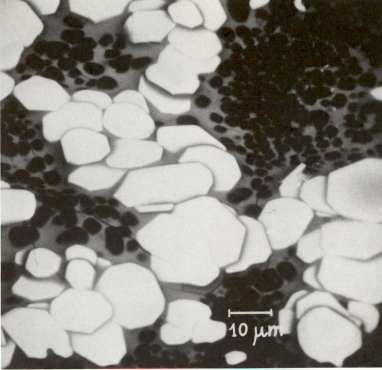

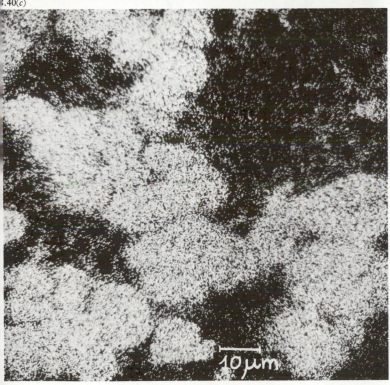

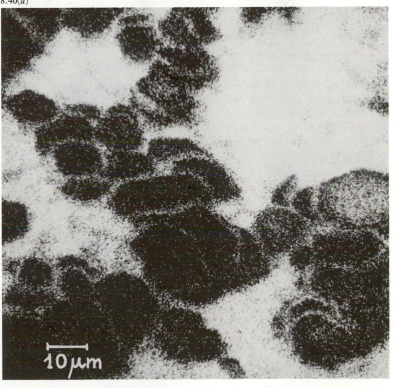

8.40 This group of scanning electron micrographs shows the same oxidised area of a silicon nitride ceramic viewed by four different imaging and analysis techniques. Figure 8.40(a) is a conventional secondary electron image which shows that the oxidised surface consists of polygonal plates set in a somewhat granular background matrix, but it gives no information about the composition of these structures. Figure 8.40(b) is an image formed from back-scattered electrons, and it tells the microscopist that the brightest areas – the white polygons – contain an element of a high atomic number. This is because heavier elements back-scatter more electrons.

Figure 8.40(c) is a map of X-rays emitted at the wavelength for the metal cerium, and it shows that the polygonal plates consist largely of cerium oxide, which was used in making the ceramic. Figure 8.40(d) is a second X-ray map, at the wavelength for silicon; it shows that the granular background matrix is predominantly silicon. X-ray mapping can be used to identify all elements with an atomic weight greater than that of fluorine.

SEMs; 8.40(a) secondary electron mode; 8.40(b) back-scattered mode; 8.40(c) cerium X-ray map; 8.40(d) silicon X-ray map; all ×1000

CHAPTER 9
EVERYDAY WORLD

IN this final chapter, the subjects are close to home: familiar items seen in unfamiliar close-up. The pictures illustrate well the fascination which microscopes hold for those lucky enough to use them. A single image can reveal the structure or workings of an object which would take hundreds of words to describe with equal clarity. The most mundane objects can take on a startling or striking appearance.

9.1 This picture is of the most ordinary subject it is possible to imagine – house dust. The specimen was obtained simply by tipping out the contents of a domestic vacuum cleaner. House dust consists of fragments of soil brought in on the soles of feet or shoes, fibres which have escaped from clothing or hair, and skin scales from the inhabitants of the house. Not all the components of dust are dead, however; in this scanning electron micrograph, the object on the right-hand side of the picture is a tiny dust mite, a species of *Glycyphagus*. All vacuum cleaners contain dust mites, and so do all carpets and mattresses. Each gram of dust contains about 1000 mites. They spend their days perambulating a landscape of, to them, huge rocks and tangled jungles. They eat the human skin scales with which we fill their microscopic world. This one is moving from right to left. The boulder it is about to encounter is a tiny grain of sand.
SEM, ×145

FABRICS

Fabrics are produced from a variety of fibres, both natural and man-made. Natural fibres come from plants (cotton, flax), animals (wools), or insects (silk). With the exception of silk, natural fibres are usually short; they are also very fine, and must be spun together to produce a thread of useful length and thickness. Man-made fibres, on the other hand, are extruded continuously from dies; their length and thickness is controlled during manufacture. The fabric itself is made by weaving or knotting threads together.

9.2 Lace-making is a skilled craft; in this micrograph of a piece of machine-made lace net, each cell of the net can be seen to be a complex series of knots. The thread is a multi-stranded cotton and polyester yarn.
SEM, ×7

9.3 This fabric is machine-made from an artificial fibre. It is a piece of net made from polythene. Each cell of the net is identical to its neighbours, and the thread is a single extruded fibre of plastic. The micrograph was made using polarised light, and this accounts for the colours, which correspond to residual stresses in the plastic material of the fibre.
LM, polarised light, ×35

9.2

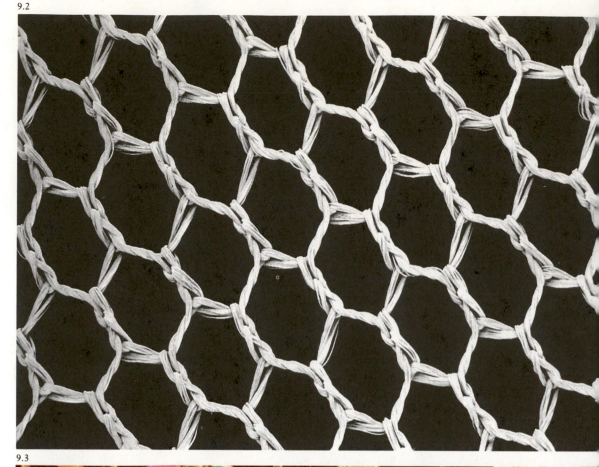

9.3

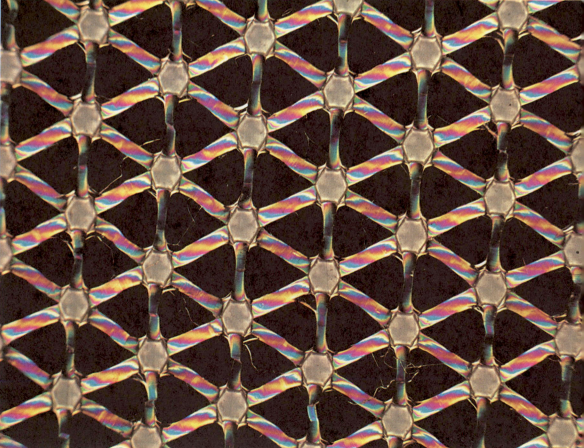

9.4–9.5 This pair of scanning electron micrographs shows the difference between a clean and dirty shirt collar. In Figure 9.4, the clean collar is seen to be made of woven cotton threads, each of which consists of many individual cotton fibres spun together. Figure 9.5 reveals the transformation wrought by just one day's city wear. The fabric has become encrusted with a layer of grease, sweat and particles of dust, together with dead scales from the wearer's skin. Happily, washing the fabric in detergent will restore the pristine appearance shown in Figure 9.4.
SEMs, both ×100

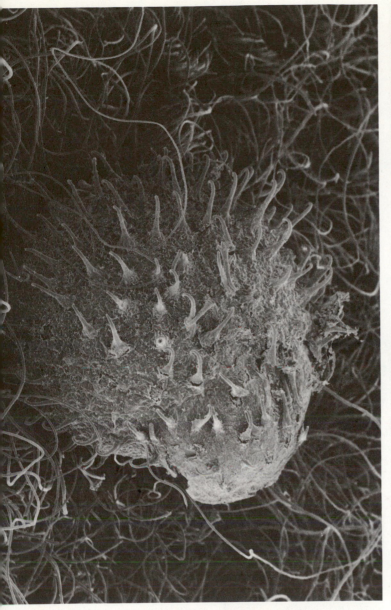

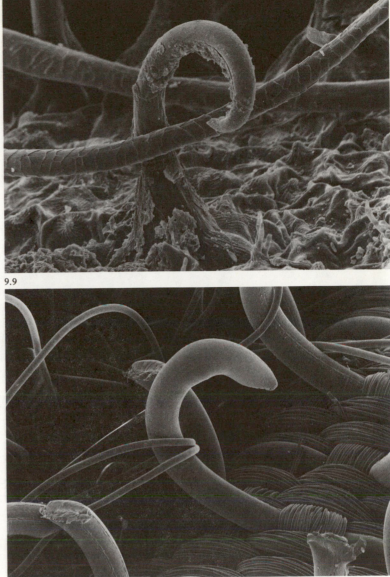

9.9

VELCRO

One day in 1955, the Swiss inventor George de Mestral was out hunting in the Alps with his dog. Both became covered with burrs from the burdock plant. As he picked the offending objects from his clothing and from his dog's coat, Monsieur de Mestral had a bright idea. Why not design a fastener using the same principle? Velcro had arrived.

Velcro is made from nylon and it comes in two parts – one side consisting of hooks and the other side consisting of loops. When the two parts are pressed together, the hooks engage the loops, and the fastening is made. Velcro is strong enough to resist a direct pull, but is easily dismantled by a peeling action. Its particular virtue is its ease of use, even when manual skill is impaired through injury or by wearing gloves. Velcro fasteners have been to the top of Everest, and out into space. It is also a durable fastening, able to withstand laundry, and able to work after 5000 cycles of opening and closing.

9.6 This photograph shows the two parts of Velcro in the separated state. The hooks (top) consist of loops of thick nylon which have been snipped open during manufacture. The loops (bottom) are closed, and are made from multiple strands of thinner nylon which can pass through the gaps in the hooks.
Macrophotograph, ×85

9.7 Plants have been using the idea of Velcro for millions of years. This scanning electron micrograph shows a fruit of the goosegrass, *Galium aparine*, attached to a woollen sweater. The loose fibres on the surface of the garment have entangled themselves in the projecting hooks on the surface of the fruit.
SEM, ×20

9.8 At higher magnification, the mechanism is clearer. Each hook is beautifully curved and finishes in a sharp point. In this picture, one hook has captured a single fibre of wool; another fibre is in the background. The scales on the surface of the wool are characteristic of animal hair; in this case, that of sheep.
SEM, ×250

9.9 The equivalent view of Velcro shows that the stout hooks are woven into a base of fine nylon material for flexibility. Nylon is extruded from fine dies, and shows no surface texture. The hook in this picture is about eight times larger than the natural one shown in Figure 9.8.
SEM, ×30

9.10

9.11

9.12

9.13

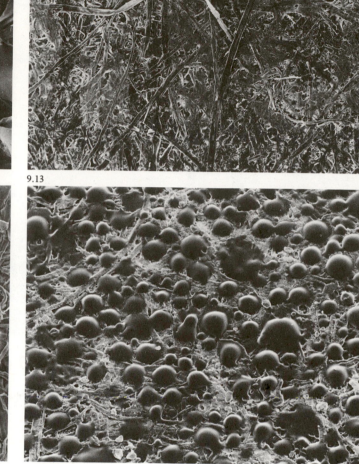

PAPER

Most paper is made from cellulose fibres extracted from wood. Other fibre sources such as rags, flax and hemp are used to make fine quality papers used for legal documents or cigarette wrappings. In the production process, the fibre source – wood chips, cleaned rags, the stems of flax, jute, or bamboo – is beaten in water to make a pulp. The liquid pulp is treated with chemicals to remove impurities, bleached, and filtered on a woven screen, producing a wet mass of tangled fibres which is then rolled, compressed and dried to yield the sheet of paper.

Paper varies enormously in quality, according to its thickness, density and fibre source. The cellulose fibres from softwoods (conifers) are longer than those from hardwoods; softwood paper is stronger, but hardwood paper has greater opacity, and can be made smoother. Cotton fibres from rags are very long and low in impurities; the paper they produce has great durability and quality, and is used for banknotes, tracing paper and carbon paper.

9.10 The paper base of sandpaper is a high-quality product made from hemp or jute fibres. In this scanning electron micrograph, the paper itself is not visible. The picture shows the particles of abrasive, consisting of ground glass, which are bonded to the surface of the sheet. The specimen is a coarse grade of sandpaper; finer grades use smaller glass particles.
SEM, ×50

9.11 Newsprint is made from sawn logs ground between abrasive stones. The pulp contains all the impurities of the original timber: broken fibres, lignin and other cell wall components. Paper from this type of pulp is never very white, and discolours easily. It is opaque, however, and accepts printing inks well. The specimen is from a page of *The Times* newspaper, and shows the letters 'nd' – from the word 'London' – printed on it.
SEM, ×60

9.12 Viewed at roughly the same magnification, soft toilet tissue has a much more open structure, with fewer impurities and longer, wider fibres. It is made from pulp which has been beaten and chemically refined. As a result, it is very absorbent.
SEM, ×50

9.13 Each of the round objects in this micrograph is a tiny resin sphere filled with glue and incorporated into the surface of a sheet of paper. When the sheet is pressed onto a firm surface, it becomes stuck as some of the spheres burst and release their adhesive. The process can be repeated many times, since only a few of the spheres burst on each occasion. The picture is of a Post-It sticker.
SEM, ×70

WATCH

9.14 The Swiss watch epitomises precision engineering in most people's minds. It is a reputation that is well deserved, as this scanning electron micrograph shows. Even when substantially magnified, the mechanism presents a picture of precision; the slightly off-centre appearance of the middle gear wheel is in fact due to the specimen being tilted on the microscope stage. The picture is of the crown wheel of a 17-jewel Incabloc movement. When the watch is wound, the winding pinion (the small gear wheel seen edge-on at bottom centre) rotates, meshing with the crown wheel. The rotation of the crown wheel causes the ratchet wheel at top left to turn, and this wheel is attached to the barrel which holds the mainspring of the watch. The mechanism is very clean; the very small particles of dust on the surface and on some of the teeth of the crown wheel would not affect its function at all. The watch had never been serviced since its assembly, and this is reflected in the absence of screwdriver marks in the slotted pillar of the crown wheel.

SEM, ×16

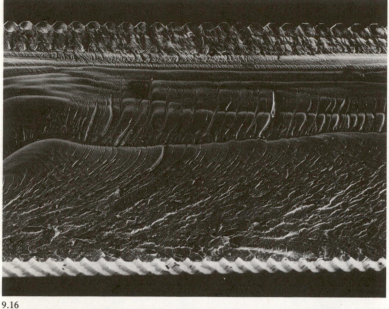

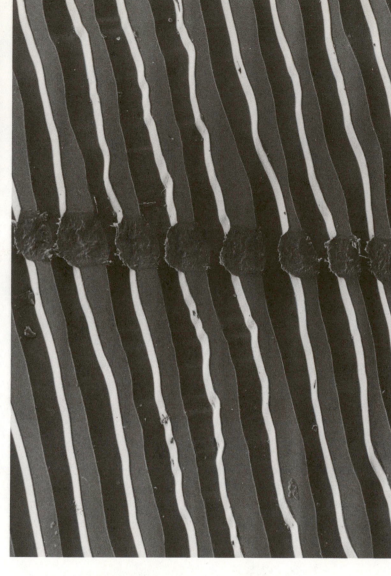

9.16

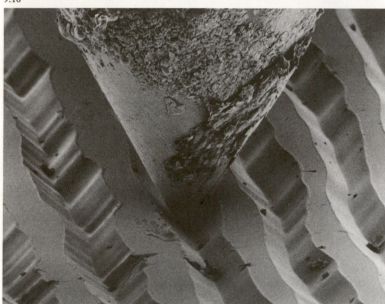

RECORDS & DISCS

A gramophone record is a remarkable object, even though a very familiar one. Within its plastic surface is stored all the information to reproduce 30–40 minutes of the sound made by an orchestra consisting of dozens of instruments. Microscopy shows how this is achieved, and in particular it emphasises the difference between the old-fashioned long-playing record and its modern counterpart, the compact disc.

9.15 An LP has two 'sides'. Each consists of a modulated spiral groove pressed into the surface of the plastic disc. In this scanning electron micrograph of a cross-section through a broken LP, the grooves appear as the saw-tooth patterning towards the top and bottom of the picture. The markings on the main body of the disc are fracture patterns in the PVC from which the record is made.
SEM, ×30

9.16 Each V-shaped groove on the LP's surface has undulating side walls. The stylus follows the contours of these waves, and the way in which it moves determines the sound which is produced by the loudspeakers. In a loud passage, for example, the waves on the sides of the groove are deep; in a quiet passage, they are shallow. A high note is recorded as a rapidly changing wave, whereas a note of low pitch is recorded as a slowly changing wave. In this micrograph, a diamond stylus sits in the groove. Before making the micrograph, the stylus was carefully cleaned; nonetheless, the picture shows that it is very dirty. The small particle in front of the stylus is a piece of dust. As it passes over the dust, the stylus will be rapidly jerked upwards and this will give rise to a high-pitched click in the loudspeakers. The part of the groove in this picture represents a loud orchestral passage in the *Sinfonietta* by Janacek.
SEM, ×90

9.17 A major drawback with LPs is the ease with which they can become scratched. This micrograph shows the effect of accidentally drawing a fingernail across the surface of an LP. The resulting scratch narrows the top of each turn of the groove, and causes a sudden movement of the passing stylus. The result is an annoying, loud click each time the disc revolves – every 1.8 seconds.
SEM, ×60

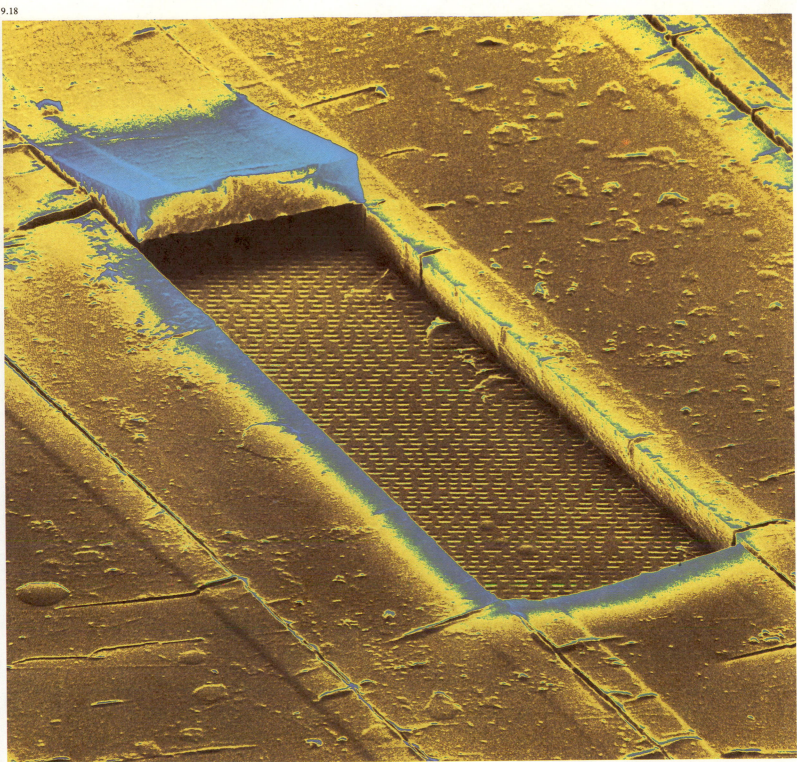

9.18 The compact disc works in a completely different way to the LP. One side of the disc is pressed to produce a pattern of very small elongated bumps, laid out in the form of a fine, continuous spiral which, in a typical CD, is 20 km long. The turns of the spiral are so close together that 60 of them would fit into the width of a single groove of an LP. The size of each bump and the rate at which adjacent bumps change in size determine the volume and pitch of the sound produced. The bumps are not touched by a stylus. Instead, they are coated with a thin layer of aluminium, which reflects the light from a laser beam focused on the metalised surface. As the disc rotates, the light is reflected to a sensor as a stream of rapid flashes, and this signal is processed to produce the music in the loudspeakers. The delicate surface of the disc is protected by a layer of transparent plastic. In this picture, the protective layer has been cracked open and partly removed. Beneath the protective layer, the pattern of bumps is visible – part of the first movement of Mozart's 40th symphony. The bumps vary in length from 0.83 micrometres to 3.56 micrometres. These distances are close to the wavelength of visible light. This is why shining a light onto a CD produces a rainbow-coloured reflection, by the process of diffraction.
SEM, false colour, ×1040

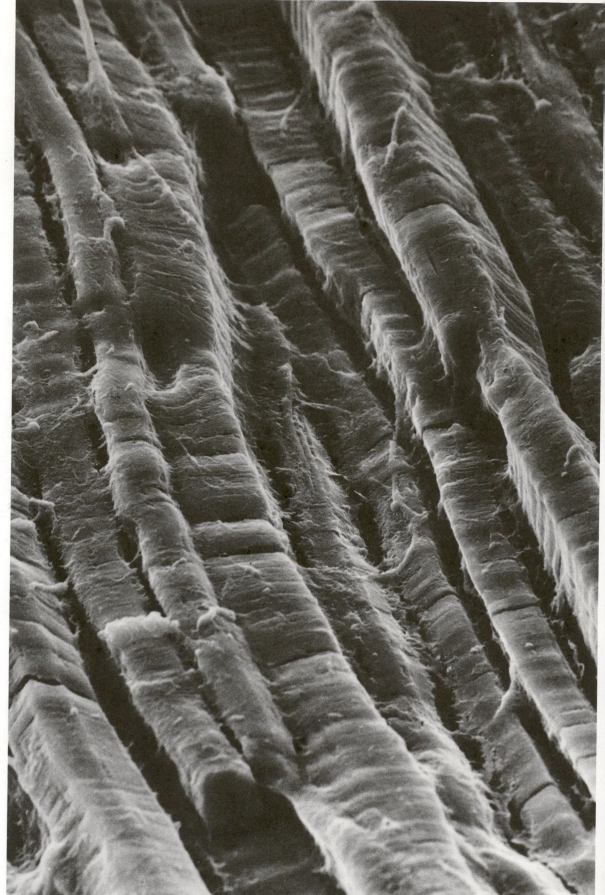

Cooking can be a chore or an art form – depending on the cook. To a microscopist, however, cooking rarely improves a raw ingredient. The exquisite architecture of cells and tissues is irreversibly destroyed as heat coagulates proteins, dissolves fats, and hydrolyses fibres and starch. These, of course, are the very changes which make food pleasant to taste and easy to digest. And most of us, after all, cook with eating in mind, not microscopy.

9.19 Meat is the muscle tissue of animals. In this piece of roast beef, the muscle fibres can be seen as the rectangular rods running from top left to bottom right. The striations along each rod correspond to the arrangement of actin and myosin filaments in the living muscle. The entire specimen is coated with a thin layer of fat released by the cooking. SEM, ×370

9.20–9.21 The potato is a rich source of energy in our diet due to the presence of starch in its cells. In the slice of a raw (living) potato in Figure 9.20, the starch appears in the form of particles called amyloplasts, groups of which can be seen looking like eggs in a nest. Other cells in the picture are not cut through the centre, and appear to contain no starch. The picture is full of the beauty and detail which typifies micrographs of living biological materials. After 20 minutes in boiling water, by contrast, the organisation of the potato has changed completely (Figure 9.21). The outline of the cells is still visible, but the starch has been converted into a bulky, glue-like mass. This is more easily digestible, but less visually pleasing. SEMs, both ×170

PICTURE CREDITS

Most of the photographs and other illustrations in this book are available from the **Science Photo Library** (SPL), 2 Blenheim Crescent, London W11 1NN (Tel: 01–727–4712).

1.1–1.4 Tony Brain/SPL
1.5 David Leah/SPL
1.6 Biophoto Associates
1.7–1.8 Jeremy Burgess/SPL
1.9 Manfred Kage/SPL

2.1 Manfred Kage/SPL
2.2 From *TISSUES AND ORGANS: A Text-Atlas of Scanning Electron Microscopy* by Richard G. Kessel & Randy H. Kardon. Copyright © 1979 W.H. Freeman & Company. Used by permission
2.3 Manfred Kage/SPL
2.4 Biophoto Associates
2.5 Tony Brain/SPL
2.6 G. Schatten/SPL
2.7 David Scharf/SPL
2.8 From *TISSUES AND ORGANS: A Text-Atlas of Scanning Electron Microscopy* by Richard G. Kessel & Randy H. Kardon. Copyright © 1979 W.H. Freeman & Company. Used by permission
2.9–2.10 CNRI/SPL
2.11 Manfred Kage/SPL
2.12–2.13 G. Bredberg/SPL
2.14 Sinclair Stammers/SPL
2.15 M.I. Walker/Science Source/SPL
2.16 Omikron/Science Source/SPL
2.17 Biophoto Associates
2.18 Manfred Kage/SPL
2.19 Biophoto Associates
2.20 Manfred Kage/SPL
2.21 D. Jacobowitz/SPL
2.22 Manfred Kage/SPL
2.23 CNRI/SPL
2.24 Eric Gravé/SPL
2.25 Don Fawcett/Science Source/SPL
2.26 Michael Abbey/Science Source/SPL
2.27 M.I. Walker/Science Source/SPL
2.28 Don Fawcett/Science Source/SPL
2.29 Kevin Fitzpatrick, Guy's Hospital Medical School/SPL
2.30 Eric Gravé/SPL
2.31 Biophoto Associates
2.32 CNRI/SPL
2.33 Tony Brain/SPL
2.34 CNRI/SPL
2.35 Biophoto Associates
2.36 Eric Gravé/SPL
2.37 J. Gennaro/Science Source/SPL
2.38 CNRI/SPL
2.39 J. James/SPL
2.40 Eric Gravé/SPL
2.41 Tony Brain/SPL
2.42 ASA Thorensen/Science Source/SPL
2.43 CNRI/SPL
2.44–2.45 From *CORPUSCLES: Atlas of Red Blood Cell Shapes* by Marcel Bessis, Springer–Verlag, 1974
2.46 A.R. Lawton/SPL
2.47 W. Villiger, Biozentrum/SPL
2.48 A. Liepins/SPL
2.49–2.50 Jeremy Burgess/SPL
2.51–2.52 Biophoto Associates
2.53–2.54 Jeremy Burgess/SPL

3.1 David Scharf/SPL
3.2–3.3 Eric Gravé/SPL
3.4 Biophoto Associates
3.5 Biology Media//Science Source/SPL
3.6 John Walsh/SPL
3.7–3.10 Biophoto Associates
3.11 Cath Wadforth, University of Hull/SPL
3.12–3.13 Kevin Fitzpatrick, Guy's Hospital Medical School/SPL

3.14–3.16 Sinclair Stammers/SPL
3.17–3.20 John Walsh/SPL
3.21 Jeremy Burgess/SPL
3.22 David Scharf/SPL
3.23 Biophoto Associates
3.24 Cath Wadforth, University of Hull/SPL
3.25 Tony Brain/SPL
3.26 Biophoto Associates
3.27–3.30 David Scharf/SPL
3.31–3.32 Jeremy Burgess/SPL
3.33 John Walsh/SPL
3.34–3.38 Jeremy Burgess/SPL
3.39 David Scharf/SPL
3.40 John Walsh/SPL

4.1–4.3 Jeremy Burgess/SPL
4.4 Patrick Lynch/Science Source/SPL
4.5 Chuck Brown Science Source/SPL
4.6 Jeremy Burgess/SPL
4.7–4.8 James Bell/SPL
4.9 Jeremy Burgess/SPL
4.10 M.I. Walker/Science Source/SPL
4.11–4.12 Jeremy Burgess/SPL
4.13 James Bell/SPL
4.14 David Scharf//SPL
4.15–4.22 Jeremy Burgess/SPL
4.23 M.I. Walker/Science Source/SPL
4.24–4.26 Jeremy Burgess/SPL
4.27–4.28 David Scharf/SPL
4.29 Tony Brain/SPL
4.30 Jeremy Burgess/SPL
4.31 R.E. Litchfield/SPL
4.32–4.33 Jeremy Burgess/SPL
4.34 R.E. Litchfield/SPL
4.35 Jeremy Burgess/SPL
4.36 Gene Cox/SPL
4.37–4.40 Jeremy Burgess/SPL
4.41 Biophoto Associates

5.1 Tektoff-RM, CNRI/SPL
5.2 M. Wurtz, Biozentrum/SPL
5.3 Tektoff-RM, CNRI/SPL
5.4 M. Wurtz, Biozentrum/SPL
5.5 Samuel Dales/SPL
5.6 Luc Montagnier, Institut Pasteur, CNRI/SPL
5.7 B. Heggeler, Biozentrum/SPL
5.8 Lee Simon/SPL
5.9 Biozentrum/SPL
5.10 Tony Brain/SPL
5.11 L. Caro/SPL
5.12 Eric Gravé/SPL
5.13–5.14 CNRI/SPL
5.15 Tony Brain/SPL
5.16 CNRI/SPL
5.17 Gopal Murti/SPL
5.18 Jeremy Burgess/SPL
5.19 John Innes Institute/SPL
5.20 Jeremy Burgess/SPL
5.21 Biophoto Associates
5.22 R.B. Taylor/SPL
5.23 Michael Abbey/Science Source/SPL
5.24–5.25 James Bell/SPL
5.26 Jan Hinsch/SPL
5.27–5.28 Biophoto Associates
5.29 Ann Smith/SPL
5.30 Biophoto Associates
5.31 Jeremy Burgess/SPL
5.32 Tony Brain/SPL
5.33–5.34 Biophoto Associates
5.35 Jeremy Burgess/SPL
5.36 Biophoto Associates
5.37 Jeremy Burgess/SPL

6.1 Jeremy Burgess/SPL
6.2 CNRI/SPL
6.3 Jeremy Burgess/SPL
6.4–6.5 Don Fawcett/Science Source/SPL
6.6 Don Fawcett & D. Phillips/Science Source/SPL
6.7–6.8 Don Fawcett/Science Source/SPL

6.9 Don Fawcett & D. Friend/Science Source/SPL
6.10 Don Fawcett/SPL
6.11 Don Fawcett & T. Kuwabara/Science Source/SPL
6.12 Jeremy Burgess/SPL
6.13 K.R. Miller/SPL
6.14 EM Unit, British Museum (Natural History)
6.15 Jeremy Burgess/SPL
6.16 Gopal Murti/SPL
6.17 P. Dawson, John Innes Institute
6.18 Don Fawcett/Science Source/SPL
6.19 Don Fawcett & D. Phillips/Science Source/SPL
6.20–6.21 Biophoto Associates
6.22 Don Fawcett/Science Source/SPL
6.23 Eric Gravé/SPL
6.24 Keith Porter/SPL
6.25 J. Pickett-Heaps/SPL

7.1 Jeremy Burgess/SPL
7.2 Mitsuo Ohtsuki/SPL
7.3 H. Hashimoto, Osaka University
7.4 Y.P. Lin & J.W. Steed, University of Bristol
7.5 I. Baker
7.6 Mike McNamee, Chloride Silent Power Ltd/SPL
7.7 G. Müller, Struers GmbH
7.8 C. Hammond, The University of Leeds
7.9 John P. Pollinger & Gary L. Messing, Ceramic Science Section, Department of Materials Science, The Pennsylvania State University
7.10 St. John & Logan, *J. Crystal Growth* 46, 1979
7.11–7.12 Elizabeth Leistner
7.13 Mike McNamee, Chloride Silent Power Ltd/SPL
7.14 Courtesy of Dr Riedl, Professor Jeglitsch and Dr Locker
7.15 Mike McNamee, Chloride Silent Power Ltd/SPL
7.16 Sydney Moulds/SPL
7.17–7.18 Jeremy Burgess/SPL
7.19 David Parker/SPL
7.20 Jan Hinsch/SPL
7.21–7.28 Mike McNamee, Chloride Silent Power Ltd/SPL
7.29 Lou Macchi, Poroperm-Geochem Ltd
7.30–7.31 Peter Borman, Poroperm-Geochem Ltd
7.32 Jan Hinsch/SPL
7.33–7.36 Mike McNamee, Chloride Silent Power Ltd/SPL
7.37–7.38 G. Müller, Struers GmbH
7.39 Mike McNamee, Chloride Silent Power Ltd/SPL
7.40 G. Müller, Struers GmbH
7.41 Manfred Kage/SPL
7.42–7.43 G. Müller, Struers GmbH

8.1 David Scharf/SPL
8.2–8.5 G. Müller, Struers GmbH
8.6 The Welding Institute, Cambridge Instruments
8.7–8.8 G. Müller, Struers GmbH
8.9 J.D. Williams, Queens University of Belfast
8.10 Max-Planck-Institut fur Metallforschung
8.11 G. Müller, Struers GmbH
8.12 Elizabeth Leistner
8.13 G. Müller, Struers GmbH
8.14 Elizabeth Weidmann, Struers Inc.
8.15 Courtesy of G.N. Babini, A. Bellosi, P. Vincenzini, *J. Mat. Sci.*, 19, 3, 1984
8.16 National Physical Laboratory, Crown Copyright Reserved
8.17 G. Müller, Struers GmbH
8.18 National Physical Laboratory, Crown Copyright Reserved
8.19 G. Müller, Struers GmbH
8.20 David Parker/SPL
8.21 STC/A. Sternberg/SPL
8.22–8.24 Jeremy Burgess/SPL
8.25 VG Semicon/SPL

8.26 Mike McNamee, Chloride Silent Power Ltd/SPL
8.27 Jan Hinsch/SPL
8.28 G. Müller, Struers GmbH
8.29–8.30 Jeremy Burgess/SPL
8.31–8.35 G. Müller, Struers GmbH
8.36 C.E. Price, Oklahoma State University
8.37 J.G. Ashurst, Chloride Silent Power Ltd
8.38 Shell, Thornton Research Centre
8.39 Manfred Kage/SPL
8.40 Courtesy of G.N. Babini, A. Bellosi, P. Vincenzini, *J. Mat. Sci.*, 19, 3, 1984

9.1 Jeremy Burgess/SPL
9.2 R.E. Litchfield/SPL
9.3 Harold Rose/SPL
9.4–9.5 Jeremy Burgess/SPL
9.6 Manfred Kage/SPL
9.7–9.21 Jeremy Burgess/SPL

Page 186 (all pictures) SPL
P. 187 (far & centre left) SPL
P. 187 (centre right) Neil Hyslop
P. 187 (far right) Museum of the History of Science, University of Oxford
P. 188 (left) SPL
P. 188 (centre right) Museum of the History of Science, University of Oxford
P. 188 (far right) GECO UK Ltd/SPL
P. 189 (far left & upper centre) Courtesy of E. Ruska, with thanks to T. Mulvey
P. 189 (lower centre) D. McMullan/SPL
P. 190 (left) D. McMullan/SPL
P. 190 (right) Neil Hyslop
P. 191 Neil Hyslop
P. 192 (upper left) S. Stammers/SPL
P. 192 (upper right, lower left & right) J. Patterson/SPL
P. 193 (upper & lower left) Neil Hyslop
P. 193 (far right) Jan Hinsch/SPL
P. 194 James Stevenson/SPL
P. 195 (upper left) John Durham/SPL
P. 195 (lower left) R. King/SPL
P. 196 Neil Hyslop
P. 197 (left & right) Neil Hyslop
P. 198 (upper left) Heather Davies/SPL
P. 198 (centre right) Muriel Lipman/SPL
P. 198 (far right) Neil Hyslop
P. 199 (far left) Kenneth R. Miller/SPL
P. 199 (upper & lower right) Jeremy Burgess/SPL
P. 200 Norman Costa & Sinclair Stammers/SPL
P. 201 J.G. White, W.B. Amos & M. Fordham (1987) An evaluation of confocal versus conventional imaging of biological structures of fluorescence light microscopy, *J. Cell Biol.* (in press)
P. 202 (left) Lawrence Berkeley Laboratory/SPL
P. 202 (right) Mitsuo Ohtsuki/SPL
P. 203 (right) Neil Hyslop
P. 203 (left) David Parker/SPL
P. 204 (both) Courtesy of IBM

Pictures credited to Dr Jeremy Burgess were taken during his time of employment at the John Innes Institute, Norwich.

INDEX

Acanthrocirrus retrirostris, 48
Acer saccharum, 71
acoustic microscopy, 148, 202
adenosine triphosphate (ATP), 114–15, 119–20
adenovirus, 91
AIDS (Acquired Immune Deficiency Syndrome), 91
akinete, 100–1
Alaria mustelae, 48
albite, 138–9
algae, 89, 100–1, 116–17
    blue-green, 100–1
    filamentous, 100–1
    green, 100–1
    single-celled, 102–3
alloys, 125, 131–3, 135, 145, 155, 168–70
alumina ceramic, 128
aluminium alloys, 131–3, 135, 145, 168
alveolus, 32–3
Amoeba, 45
amoebic dysentery, 44
amyloplast, 116, 184–5
anaphase, 122–4
angiosperm, 84
annual ring, 71
Anobium punctatum, 57
Anopheles, 54
anther, 78–80
antibiotic, 115
antibody, 39
antigen, 39
Antirrhinum majus, 78, 86
aphid, 42–3
Apis mellifera, 60
'Arkose', 139
arthropods, 42
Aschelminthes, 52
Astralloy, 127
athlete's foot, 105
atom, 126 *et passim*
ATP, 114–15, 119–20
attack mechanisms, 76–7
Aucuba japonica, 117
austenite, 130, 144–5
auxospore, 102
axon, 24–5, 27

bacteria, 89, 92–9, 115
    nitrogen-fixing, 98–9
bacteriophage, 92–4
bacteroid, 98–9
Balanus balanoides, 48
Barbulanympha ufalula, 44
barnacle, 48
basidia, 106
basilar membrane, 20–1
bat, 113, 115
Batrachoseps attenuatus, 120–1
Beet Necrotic Yellow Vein, 91
Bellis perennis, 80–1
Biddulphia, 102–3
Bilbergia nutans, 82–3
bilharzia, 50
binary fission, 95
bioengineering, 148, 158–9
biological specimens, preparation of, 198–200
biotite, 138–9
birefringence, process of, 11
black garden ant, 55
blood, 36–9
    human, 36–8
    proteins, C1–C9, 39
Bombylius major, 119
bone, 30–1
brass, 128–9
Brassica campestris, 67–8, 72, 85
Bravais, Auguste, 125
Bravais lattices, 124–5
    cubic, 124–5, 130

    monoclinic, 125, 157
    orthorhombic, 125
    parallelepipeds, 125
    tetragonal, 125, 157
    triclinic, 125
brazing, 150
brewer's yeast, 106
brittle fracturing, 170
bronchiole, 32–3
bronchus, 32
buttercup, 68–9

C1–C9, blood proteins, 39
cadmium sulphide, 165
calcite, 141
calcium oxalate, 75
calcium sulphate, 135–6
cambium, 70–1
cancer, 32, 39, 91
Cannabis sativa, 74–5
Cape sundew, 76
Capsella bursa-pastoris, 80–1
capsid, 90
capsomere, 90–1
carborundum, 156
carnivorous plant, 76
carpel, 78–83
carrot, 116
caterpillar hatchery, 62–3
cell, 12 (structure of) *et passim*
    blood, 36–8
    bone-building, 30
    cambium, 70–1
    cancer, 39
    collenchyma, 70
    cortical, 70
    enterocyte, 34–5
    epidermal, 40–1
    epithelial, 32–3
    eukaryotic, 89, 100–1, 108, 114–15
    eye, 60–1
    guard, 73
    hair, 20–1
    mesophyll, 72
    Müller, 19
    muscle, 28, 119
    nerve, 24–7
    osteoblast, 31
    osteoclast, 30
    parenchyma, 70
    phloem, 68
    photoreceptive, 115
    plasma, 39
    prokaryotic, 89, 108, 117
    sex, 14 *et seq.*, 82
    statocyte, 68
    unit, 125
    xylem, 68
cell process, of nerve cell, 24
cellulose fibre, 180
cementite, 142
central nervous system (CNS), 24
ceramics, 156–7
Ceratitis capitata, 55
chickweed, 78–81
'Chinese script', 147
chitin, 54, 62
Chlamydomonas asymmetrica, 10
chlorite, 138–9
chlorophyll, 100–1, 116–17
chloroplast, 72–3, 100–1, 108, 116–17, 119
cholera, 94
chromatid, 122–4
chromatin, 108–12, 120–1
chromium, 144–5
    alloy, 147
chromoplast, 116
chromosome, 12, 16, 111–12, 122–4
ciliary body, 19

cilium, 120–1
Clonorchis sinensis, 49
CNS, 24
coating technology, 148
cochlea, 20
cockroach, 44
cocksfoot grass, 82–3
collenchyma, 70
colouring, of electron micrograph, 11
common cold, 91
common salt, 124
compact disc, 182–3
cones, photoreceptive, 115
conidia, 106
conidiophores, 106
conifer, 84
conjugation in bacteria, 94
copper, 169
Coprinus disseminatus, 106
Cornish granite, 138–9
corrosion, 148, 166–7
Corti, organ of, 20–1
Cosmos bipinnatus, 78–9
cotton fibre, 180
crab louse, 57
cristae, 114–15
cristobalite, 157
crystal structures, 134–7
crystal zoning, 138–9
crystallography, 125, 128
Crytocercus puntulatum, 44
cucumber, 82
Cucurbita pepo, 82
cuticle, 72–3
Cyclotella meneghiniana, 102–3
Cylindrospermum, 100–1
cystolith, 74–5
cytoplasm, 44, 73, 94, 108–14, 118, 120
cytoskeleton, 118

Dactylis glomerata, 82–3
daisy, 80–1
Darwin, Charles, 67
daughter cell, 122–4
Deiter cell, 21
de Mestral, George, 179
dendrite, 24, 150–1, 155
    metal, 150–1, 155
    nerve cell, 24
dendritic structures, 131–3, 135
dense fibre, 120
dermis, 40–1
desmid, 100–1
diagenesis, 140–1
diamond-backed moth, 58–9
diatom, 102–3
DIC (differential interference contrast),
    see Nomarski DIC
dictosome, 113
Didinium nasutum, 46–7
digestion, human, 34–5
dislocations, 128–9
Distalgesic, 135
DNA, 7, 11, 89, 90–7, 111, 114, 116–17, 120
'doped' silicon, 160
dragonfly, 60–1
DRAM, 148–9
Drosera capensis, 76–7
duckweed, 67
dust, 174
dust mite, 174
dye, use in microscopy, 12
dynamic random access memory (DRAM), 148–9

Echinococcus granulosus, 51
echinocyte, 38
E-coli, 93–4
Eleagnus pungens, 75

electron, 7–11
    primary, 9
    secondary, 9
electron beam lithography, 148–9
electron lens, 9
electron microscopy, technical details, 196–200
electronic chip, 125
electronics, 160–5
embryo, 84–5
embryo-sac, 84–5
enamel coating, 167
endocrine gland, 15
endomembrane system, 108
endoplasmic reticulum, 113, 119–20
endosperm, 84
enkephalin, 25
Entamoeba hystolytica, 44
enterocyte, 12, 34–5
enzyme, 111
eosin, 11
epidermis, 40–1, 73
    leaf, 73
    skin, 40–1
epithelium, 32–3, 35
EPROM, 162–3
erasable programmable read-only memory (EPROM), 162–3
Erica carnea, 82–3
Erisyphe pisi, 104–5
erythrocyte, 36, 38
Escherichia coli, 93–4
etching, 125, 128–9, 131–3, 142–7, 150–2, 155–7, 169
    Beraha, 151
    chemical, 125, 128–9, 131, 145, 151–3, 156–7
    colour, 133
    deep, 133
    high temperature, 125, 128, 158
    Klemm, 151–3, 167
    Nital, 142–5
    Weck's, 155
etioplast, 117
Euglena fusca, 45
eukaryotic cell, 89, 100–1
eutectic, 131–3
extreme duty materials, 154–6
eye, 18–19, 54–61, 58–9
    compound, 54–61
    human, 18–19
    simple, 58–9

Faber, Giovanni, 186
fabrics, 176–7
failure analysis, 168–71
farina, 75
fayalith, 156
feldspar, 138–9
ferrite, 130, 142–5
ferrous metal, 142–3
fibrocartilage disc, 30
filament, 118
    actin, 118
    intermediate, 118
    micro, 118
    myosin, 118
Fleming, Alexander, 106
floret, 78–80
Floscularia ringens, 53
flower details, 78–81
fluke, 48–50
fluorescence, ultra-violet, 25, 118
flux, 156
follicle, 15, 41
follicular antrum, 15
food, 184–5
forget-me-not, 82
fractography, 168–71
Fragillaria crotonensis, 102–3

rets, 116–17
rit, 167
ruit fly, 55
rustule, 102–3
ungal disease, 104–7
ungi, 89, 104–7
uniculus, 84
usion joining processes, 150–1

gabbro, 138
Galileo, 186
Galium aparine, 86–7, 179
gallium arsenide, 148, 164–5
gamete, 14
ganglion cell, 19
garden pea, 73, 105
gastric ulcer, 35
   human, 35
   rat, 35
gastrointestinal tract, 34
gene, 90, 108, 111–12, 119
'General Sherman', 70
genome, 90
Giardia lamblia, 35
Ginkgo biloba, 67
Glycyphagus, 64, 174
goblet cell, 12, 32
Golgi apparatus, 113, 120
goosegrass, 86–7, 179
grain structures, 128–9
grain weevil, 56
gramophone record, 182–3
grana, 116–17
graphite, 143
'Greywacke', 139
gymnosperm, 84
gypsum, 135

haematite, 140
haemoglobin, 36, 38
hair, human, 40–1
H & E (haematoxylin and eosin), 11 *et passim*
Haversian canal, 31
hay fever, 82–3
hearing , human, 20–1
heart muscle, 115
heather, 82-3
hemlock tree, 71
hemp fibre, 180
hermaphrodite, 80–1
heterocyst, 100–1
Hibiscus, 80–1
high voltage electron microscopy, 201
hip replacement, 158
histology, 12
histopathology, 12
history of microscopy, 186–7
hoary mullein, 75
honeybee, 60
Hooke, Robert, 108, 186
hover fly, 60
HREM (high-resolution electron microscope),
   7–11 (definition of) *et passim*
hyatidosis, 51
Hydrarachna, 64–5
Hydrodictyon, 100–1
hymenium, 106
hypha, 104–7

igneous rock, 138–9
illumination technique, 192–4 *et passim*
   bright field, 11, 68–9, 72, 78–9, 95, 100–1, 123
   cathodoluminescence, 141
   contrast, 165, 172
   convergent beam diffraction, 127
   cross-polarised light, 134–9
   dark field, 11, 50, 53, 100–1
   filtered vertical, 60
   polarised light, 11, 71, 85, 128–9, 134–41,
     145–6, 158–9, 168, 176
   reflected light, 9, 125
   replica technique, 96–7, 111, 120–1
   Rheinberg, 11, 52, 64–5, 100–1
   transmitted light, 9, 125
   ultraviolet fluorescence, 25, 39, 118
   unpolarised light, 140–1
immune system, human, 39
immunofluoresence microscopy, 25, 39
immunoglobulin, 39

Indian lotus, 86
influenza, 90
inorganic specimens, preparation of, 200
insects, 54–9
insulin, 96
integument, 84–6
interferon, 96
interphase, 122–4
iron, 142–3
   cast, 143
   spheroidal cast, 143
iron carbide, 142

Jansen, Zacharias, 186
jejunum, 12, 34
joining of materials, 150–3
   fusion, 150–1
   solid phase, 152–3
jute fibre, 180

Kalanchoe blossfeldiana, 86
kelp, 89
keratin, 40–1, 105
kidney, 11, 120–1
Kirkuchi pattern, 127, 155
knee-jerk reflex, 24–5

Lamium album, 70
large white butterfly, 62
Lasius niger, 55
laurel, 117
leaves, 72–5
Leeuwenhoek, Antoni van, 7, 12, 89, 186
legionnaire's disease, 94
legumes, 98–9
Leptospira, 95
leucocyte, 36
light microscope (LM), 7–11 *et-passim*
lignin, 70
Ligustrum vulgare, 72
limestone, 141
liver, 120
lumen, 113
lymphocyte, 39, 108–9

macrophage, 38–9
magma, 138
magnification, definition of, 8
maidenhair tree, 67
maize, 108, 117
malaria, 44, 54
marijuana, 74–5
martensite, 130
material specimens, preparation of, 200
meat, 184
meiosis, 14
meristem, 108
mesophyll, 72
mesosome, 94
metamorphic rock, 138–9
metamorphosis, 54
metaphase, 122–4
Micrasterias, 100–1
microchip, 160–5
micrograph, 8–11
microhardness measurement, 172
micrometre, definition of, 8
micro-organism, 88–107
microscope types, 7–11 (definitions), 163, 194
    et passim
   differential interference contrast, DIC
    (Nomarski), 85, 100–2, 142–7, 150, 165, 169
   light (LM), 7–11
   scanning acoustic (SAM), 163
   scanning electron (SEM), 7–11
   scanning transmission electron (STEM), 7–11
   transmission electron (TEM), 7–11
   ultraviolet fluorescence, 118
microtubule, 118, 120
microvilli, 120–1
milkweed, 76
millipede, 8
mites, 64–5
mitochondria, 89, 108–9, 113–17, 120
mitosis, 14, 122–4
mitotic spindle, 122–4
monoclinic crystals, 135–7
moth fly, 58–9

mouse spleen, 39
mucosa, 32, 35
   gastric, 35
   respiratory tract, 32
Müller cell, 19
multipolar nerve cell, 25
muscle, human, 28–9, 33, 115
   cardiac, 28, 115
   skeletal, 28
   smooth, 28, 33
   striated, 28
muscovite, 140
Muscularis mucosae, 12
mushroom, 104–7
mycelium, 104–7
mycoplasma, 88–9
Mycotypha africana, 105
myofibril, 28
Myosotis alpestris, 82
Mytilina, 53

nanometre, 8, 10, 126
Navicula monilifera, 102–3
nectar, 83
Nelumbo nucifera, 86
nervous system, 24–7
neurone, 24–5, 27
neurotransmitter chemical, 27
neutrophil, 36, 38
newspaper, 180
nickel, 144–5, 169
nickel–aluminium alloy, 128, 169
Nicotiana tabacum, 110, 117
nicotinic acid, 106–7
nitrogen-fixing bacteria, 98–9
Nomarski DIC, 85, 100–2, 142, 144–5, 147, 150,
   165, 169
non-ferrous metal, 144–7
nucellus, 84–5
nucleic acid, 90–3, 111–12
nucleoid, 116–17
nucleolus, 108–9, 120–1
nucleus, 10–36 *passim*, 73, 85, 89, 100–1, 108–24
   passim
nucleus pulposus, 30
nylon, 179
Nymphaea, 75, 78
   alba, 75
   citrina, 78

oestrogen, 15
olivine, 138
oocyte, 14, 15
opium poppy, 83–4
opossum, 113
organelle, 108–9, 113–20 *passim*
orthoclase, 138–9
ossicle, 20
osteoblast cell, 30–1
osteoclast cell, 31
osteocyte, 31
osteoid, 30–1
osteomalacia, 31
oval window, 20
ovule, 79, 82–6

pancreas, 34
Papaver somniferum, 83–4
paper, 180
papillae, 23
Paramecium, 46–7
parasite, 104
parenchyma, 70
parthenogenesis, 52
Passiflora caerulea, 82–3
passion flower, 82–3
pearlite, 142–3
pediculosis, 57
Penicillium, 106
   chrysogenum, 106
   notatum, 106
pericycle, 69
petal, 78–80
petrified wood, 141
petrology, 125, 138–9
phagocytosis, 38–9
phase, 125
phenocrysts, 138–9

Philodina, 52–3
   gregaria, 53
phloem, 68–70, 116–17
photoemission electron microscopy, 203
photon, 9, 11
photoreceptive cell, 115
photosynthesis, 67, 72, 100–1, 104, 108, 116
Phthirus pubis, 57
Pieris brassicae, 62
pine, 84
Pisum sativum, 73
pith, 70
plague, 94
plankton, 102
plasma, blood, 36, 39
plasma membrane, 108
Plasmodium, 44
plastid, 108, 116–17
platelet, 36, 142–3
   blood, 36
   metal, 142–3
Plutella xyphostella, 58–9
pneumonia, 89
polarised light, 11
polio, 91
polished sections, 128–59 *passim*, 167–9, 172
pollen, 78–84
pollination, 67, 76, 80–5
   cross, 80–3
   insect, 82–3
   wind, 80, 82–4
polystyrene, 170
polythene, 176
porphyritic basalt, 138–9
post-synaptic membrane, 27
potato, 76, 104, 116, 184–5
   blight, 104
powdery mildew, 105
preparation techniques, specimen, 194–5,
   198–200
Primula malacoides, 73, 75
privet, 72
progesterone, 15
prokaryotic cell, 89, 100–1, 108, 117
prophase, 122–4
protection devices, 76–7
protoplast, 110
protozoa, 42, 44–7, 89, 108, 120
Pseudomonas fluorescens, 94
Psychoda, 58–9
pubic louse, 57
Purkinje cell, 24, 27
PVC, 182–3

quantitative microscopy, 172–3
quartz, 138–40
quartzite, 139

rabies, 90–1
Rafflesia, 78
Ranunculus acris, 68–9
rat stomach, 35
redwood, 67
reflected light in metal microscopy, 125
refractory brick, 156
replica, 96–7, 111, 120–1
   freeze-fracture, 111, 120–1
   shadowed, 96–7
reproduction, human, 14–17
resolution limit, 7–9
respiration, human, 32–3
retina, human, 18–19, 115
Rhizobium leguminosarum, 98–9
riboflavin, 106–7
ribosome, 93–4, 108, 111–17
rickets, 31
ringworm, 105
RNA, 11, 90–3, 111–12
rocket motors, 154
rods, 115
roots, 68–9
rose, 80
rotifers, 52–3
roundworm, 52
Royal Society of London, 7
Ruska, Ernst, 7
Ruska, Helmut, 7
rust, 166–7
rust fungus, 105
rye, 68

Saccharomyces, 106–7
  cerevisiae, 106–7
  ellipsoideus, 106–7
salamander, 120–1
SAM (scanning acoustic microscope), 163
sandpaper, 180
sandstone, 140
saprophyte, 104
'sawtooth' structure, 142
scanning acoustic microscope (SAM), 163
scanning electron microscope (SEM), 7–11
  (definition of), 197 *et passim*
scanning optical microscopy, 201
scanning transmission electron microscope
  (STEM), 7–11 (definition of) *et passim*
scanning tunnelling microscopy, 204
Schistosoma, 48–50
  haematobium, 50
  japonicum, 48, 50
  mansoni, 50
schistosomiasis, 50
sea urchin, 16
sebaceous gland, 41
sedimentary rock, 138–9
seed plants, 66–7
seeds, 84, 86–7
SEM (scanning electron microscope), 7–11
  (definition of), 197 *et passim*
sensory receptors, 24
sepal, 78–9
Sequoiadendron giganteum, 70
Sequoia sempervirens, 67
sex cells, development of, 14–17
shepherd's purse, 80
sickle cell anaemia, 38
silicon, 125, 127, 147, 156–7, 173
  carbide, 156–7
  nitride, 157, 173
silicon crystal, 148, 160, 165
silicon microchip, 148, 160–5
Sitophilus granarius, 56
skeleton, human, 30–1
skin, human, 40–1
slag, 156
sleeping sickness, 54
slip, of atoms (definition), 127
slipped disc, 30
snapdragon, 78, 86

sodium chloride, 124–5
solar cells, 165
solid phase joining processes, 152–3
sperm, 14–17, 49, 66–7, 82–4, 120
  human, 14–17, 120
  plant, 66–7, 82–4
spermatogonia, 14, 16
spinal ganglion, 25
Spirogyra, 100–1
sporangia, 105
spore, 104–5
sporopollenin, 82
springtail, 58
stains used in microscopy
  anti-tubulin, 118
  anti-vimentin, 118
  eosin, 12, 23–4, 28, 33–4
  false, 17, 27, 36–7, 42–3, 55, 66, 88–9, 91,
    95–7, 108–9, 126, 148–9, 183
  Giemsa, 36
  gold chloride, 27
  haematoxylin, 12, 23–4, 28, 33–4
  negative technique, 16, 39, 90–5
  picro thionin, 31
  silver stain, 24, 27–8
  trichrome, 11–13, 20–1, 15, 31
  van Gieson, 34
stamen, 78–80
Staphylococcus epidermis, 95
starch grain, 117
statocyte cells, 68
steel, 125, 130, 142–5, 166–7, 170
stele, 69
Stellaria media, 78–9
stem, 70–1
STEM (scanning transmission electron
  microscope), 7–11 (definition of), 202 *et passim*
Stentor coeruleus, 45
sterocilia, 21
stigma, 78, 80–3
stinging hair, 77
stinging nettle, 77
stomata, 72–3, 75
streptococci, 94
striae, 102–3
stroma, 116–17
sugarbeet, 91
sugar maple, 71

sulphur, 134–5
sulphur dioxide pollutant, 166
sundew, 76
'superalloys', 147
surfactant, 33
suspensor, 84–5
sweat gland, 40
Swiss watch, 181
symbiosis, 98–9
synaptic cleft, 27
Syrphus ribesii, 60

Taenia solium, 49
tapeworm, 48–9, 51
taste, 22–3
  human, 23
  rabbit, 22
tectorial membrane, 20–1
telophase, 122–4
TEM (transmission electron microscope), 7–11
  (definition of), 196 *et passim*
tetrahydrocannabinol resin, 74–5
titanium alloy, 146, 158, 168
toadstool, 104–7
tobacco plant, 110, 117
Toxocara canis, 51
trachea, 32
Tradescantia, 73
transistor, 127, 148, 160–5
transmission electron microscope (TEM), 7–11
  (definition of), 196 *et passim*
transmitted light in petrology, 125
treehopper, 58–9
trichome, 72, 74–7
Trichophyton interdigitalis, 105
tsetse fly, 54
Tsuga canadensis, 71
tungsten, 154
tungsten carbide, 154
turbine blade, 155
turnip, 67–8, 72, 85

ultraviolet fluorescence microscope, 8 (definition
  of), 118
unit cell, 125
uranium acetate, 11
uranyl acetate microcrystal, 126

Uromyces fabae, 105
Urtica dioica, 77

vanadium, 168
van Leeuwenhoek, Antoni, 7, 12, 89, 186
vascular bundle, 70
velcro, 179
Verbascum pulverulentum, 75
villus, 34–5
virions, 90
virus, 89–91, 92–3, 110
vision, human, 18–19
vitamin B1, 106–7
vitamin C, 136–7
vole, 105
Volvox, 100–1

water lily, 75, 78
Weill's disease, 95
welding, 150–3, 162
  explosive, 150–3
  friction, 150–3
  fusion, 152–3
  resistance, 150–1
  solid phase, 152–3
  'spot', 150–1
  ultrasonic, 162
wheat, 68–9
white dead nettle, 70
Wilson, James, 186
Wolffia arrhiza, 67
wood, 70–1
woodworm beetle, 57
worms, 42, 48–51
  parasitic, 48–51

X-ray, 125, 172–3, 203
  microscopy, 203
xylem, 68–71

yeast, 89
yttrium, 156

Zea mays, 108, 117
zirconia, 157
zona pellucida, 15
zygote, 84